EDGED WEAPONRY
OF THE THIRD REICH

Edged Weaponry

SA-STABSCHEF LUTZE AND ITALIAN GENERAL RUSSO REVIEWING AN ARMY HONOR GU

of the Third Reich

BY MAJOR JOHN R. ANGOLIA

ERLIN, JULY 1938. NOTE WHITE-HANDLED FASCIST MVSN DAGGER WORN BY LUTZE.

1st EDITION

ISBN No. 912138-11-4

PRINTED IN THE UNITED STATES OF AMERICA

DESIGNED AND ILLUSTRATED
BY
ROGER JAMES BENDER

 R. JAMES BENDER PUBLISHING, P.O. BOX 1425, MOUNTAIN VIEW, CALIF. 94040

INTRODUCTION

When I undertook the subject of Nazi period blades almost twenty years ago, it was from a personal interest, and not with the intent of becoming an authority on the subject. "Swords of Hitler's Third Reich" was the first to deal exclusively with the subject of swords. Available information was extremely lacking, and the support rendered in compiling an authoritative work was lacking as well. At the time of its writing, the best effort was put into the reference considering my own limitations and the limited source material from which to draw factual data.

This work was followed by "Daggers, Bayonets and Fighting Knives of Hitler's Germany". Over twelve years of research went into the compilation of this reference, and I was satisfied that it was not simply a rehash of old information gathered between new covers. Every effort was made to present new and factual information derived from documented sources. Still it was recognized that gaps in available information existed, and in some cases, previous misconceptions and errors were allowed to continue in place of fact.

It is this subject...errors contained in an authoritative reference.. ..which I would like to briefly discuss. To touch on the subject will possibly shed some insight into the problems connected with writing a reference, and how these problems are often passed on to the reader. If the reader will permit me, I would like to quote from an article written by Waverly Root in the "International Herald Tribune" entitled "The Art of Muddying the Pool of Human Knowledge"...

> ..."It is necessary to be wary in dealing with the reference books; and I have the impression that it is becoming progressively more difficult to avoid their errors as we become more efficient in filing information and opening access to it; we file the misinformation too, where everybody can get to it. How often I have dredged up a morsel of fact which seemed on its face value to be unsound, but which I then found was confirmed by other authorities; and how often have I discovered in the end, sometimes because I found the same information given by different authors in exactly the same words, that what I had encountered was not confirmation, but simply repetition. Everybody is rewriting everybody else (including myself, but I try to be suspicious), so that the same error from the same source is repeated again and again until it is accepted as gospel."

Such, for example, is the error that existed between the identification of the Government and Diplomatic Official daggers. Mr. Root goes on to put his finger on one of the most perplexing problems that an author has to face. He states...

> ..."For book publishing, at least, it is inherently a vicious system. A publisher presumably engages an author to write a book on a given subject because he knows more about it than anybody else. He then turns the manuscript over to a minor assistant who knows nothing in particular about the subject to edit out its mistakes. Let us imagine the author has worked diligently to ferret out a truth misstated in the reference books and set the record straight. What the expert has written sounds strange to the uninitiated copy editor, for it conflicts with the generally accepted error. What does she do? She looks up the point in doubt in the reference books and triumphantly restores the error the author has vainly attempted to eliminate".

My thanks to Mr. Root for expressing so well what most authors have to live with. So what is the point of this discourse? The point is that I accept that errors have been made in my previous works, either by myself or during some phase of printing, and my research continues in an effort to find the facts and put the errors straight. While a degree of repitition is needed from the two previous books in an effort to provide continuity, this reference is largely compiled from all new data, and is intended to be the last reference I produce on the subject of blades.

I apologize for those errors which I have perpetrated or carried over. It is my sincere wish that this reference might serve to correct those errors, and provide the reader with additional knowledge to make his collecting more enjoyable.

ACKNOWLEDGEMENTS

The response to my request for assistance in compiling this final reference on blades was immediate and extensive. In many cases, I was loaned photographs which were dear to the individuals concerned, but with the knowledge that they would serve to better enhance this presentation. In almost every case, the input was made with an attitude of helping others, and not what personal gain could be derived from the act of giving. It is to the unselfish contributor that I owe a great deal of gratitude. Without their assistance and encouragement, this reference would never have been completed.

For years, I had been told how unresponsive the blade firms of Solingen, Germany, were when queried for information and assistance. I found just the opposite to be the truth. Granted hard copy information was sorely lacking due to its destruction during or after the war, but everyone I approached was more than generous with his time, and his willingness to discuss every aspect of war-time production and controls.

Mr. Ulrich Haering, assisted by Master of Saber, Hugo Möthe, gave me the better part of their day in touring and discussing the Eickhorn firm. Frau Zinndorf, of the Puma firm, so gratiously made available to me their last remaining war-time sales catalogue. Mssrs L. K. Lesch(Alcoso) and Carl Krebs (Krebs) took time out from their busy schedules to give me answers to many of my questions. My special thanks goes to Dr. Hanns-Ulrich Haedeke, Director of the Deutsches Klingenmuseum, Solingen, for a great deal of the original source material cited has been provided by him.

Needless to say, without the photographic expertese of Andy Southard and Andy Walker, and the translation assistance of Rosemarie Lehder, this reference would have been an impossible dream.

It is virtually impossible to establish an order of merit to those who have given so freely of their time, knowledge and encouragement. Below are but a few individuals and organizations that I would like to single out:

<u>Institutional</u>

Deutsches Klingenmuseum, Solingen, Germany
Imperial War Museum, London, England
Library of Congress, Washington, D.C.
Museum of the Corps of Royal Engineers, Brompton Barracks, England
National Archives, Washington, D.C.

<u>Individuals</u>

Kim Alstott
Lt. Gary Barnum
Doug Barton
Robert Bolton
Ivan Brouard
John Boutell
Marc Brown
John Byrd
Hugh Brock
Stan Bozich
Charles Campbell
Phil Capewell
Eric Campion
George Canaday
Cpt. R.J. Cole
John Coy
Ron Fabian
Fred Fallgren
Chris Farlowe
Hartvig Fleege
Tom Flurchick
Tom French
Albert Goosens
Ron Hickox
Maj. Tom Johnson
LeRoy Kile
Dr. K-G. Klietmann
Paul Lefort
Christer Lundblad
Roger Mace
Andrew Mollo
Graham Morris
J.W. Morton
Bob Moses
Richard Mundhenk
Walt Nichols
L.H.M. Pierotti
Larry Price
Fred Ramm
Bruce Regnemer
John Schleicher
Tom Sauers
Ken Schindler
Richard Seebach
Bob Settle
Hans Siegel
Albert Skeets
Col. C.J. Stewart
Fred Stephens
Harry Stromberg
Bob Thibodaux
Alois Tress (Hauptkommissar)
Arnold R. Tieche
LTC Joachim Uebe
Gary Walker
Dr. W.P. Wagner
Herman Ward
Ken Ward
Jerry Weiblen
John Vroblick

AUTHOR'S REQUEST

Now that I have completed the subject of Nazi-era blades to the extent of my ability, I intend to turn my attention to the subject of German orders and decorations, 1930-1945. Recognizing that a great deal has been produced on the subject, I wish to present the subject in a different and possibly more enlightening manner. It is my intent to cover the motivational aspect of orders and decorations, the historical development, and case histories of actual awards. Again I put myself at the mercy of the reader in requesting the use of photographs, regulations and documentation concerning the award, wear, etc. Full credit will be given for any material used. Research has been under way for over two years, and I am about to get fully involved with producing a worthwhile text. If you have anything at all which you feel might be of value, please send to:

Maj. John R. Angolia
HQ, EUCOM (J-2)
APO NY 09128

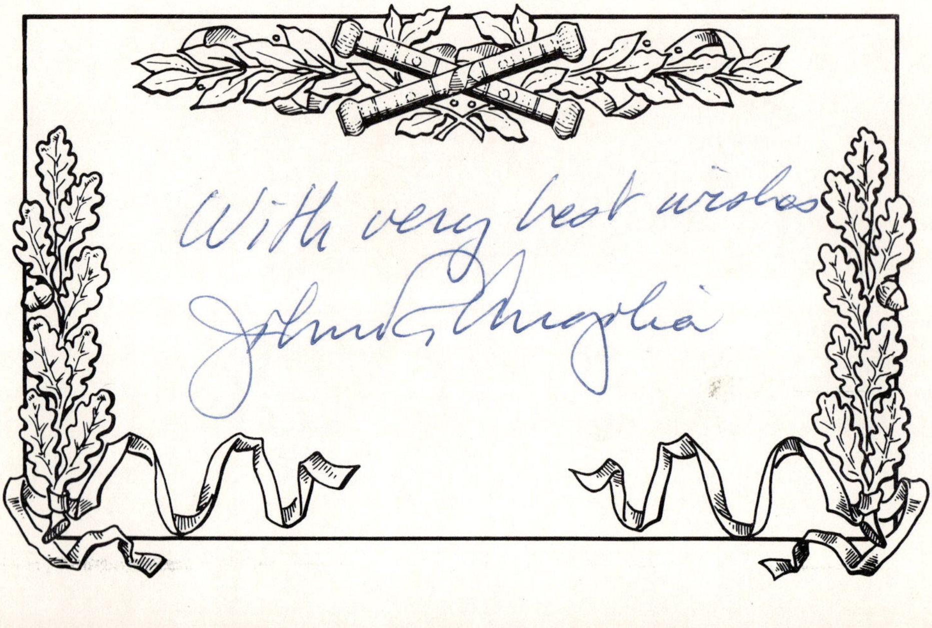

TABLE OF CONTENTS

THE CITY OF SWORDS

Upon entering the outskirts of Solingen on a narrow winding road, the first indication that you are in the heartland of the finest produced steel blades in the world is a heavy, cast-metal town marker. The unique marker is different in size, color and individuality from all others previously encountered. The town name of Solingen, of course is present, but majestically arcing under a red, blue and gold crest signifying the City of Swords.

The town is a composite of old and new, gracefully combining a farm community with thriving industry. It is more than just another picturesque German town, but a city bustling with commerce and polished with a liberal sampling of the arts. The impact of the blade trade is evident wherever one travels.... the streets named for the age-old firms, and the finished cutlery displayed in the shop windows which boast the sale of blades having the best reputation in the world.

Part of the great industrial Ruhr Valley complex, Solingen is located on the rolling high ground adjacent to the Wupper River. The narrow, but swift river at this point combines with the natural mineral resources of iron, aluminum and lead to provide the blade artisans with the necessary ingredients to produce the finely tempered blades.

Solingen (or Solagon as it was originally named) became active in European competition in the fifteenth century when blade artisans created implements beyond their own defensive and agricultural needs, and pitted their ability against the already established neighboring craftsmen from Cologne, and eventually against those from Bordeaux, Milan and even Toledo itself. Prior to this time, father had taught son how to produce the implements for survival to meet primarily the needs of the immediate family and those of the land-holder. The skills were passed on from generation to generation without a great deal of refinement. The Solingen craftsmen, however, became noted throughout Europe for the artisans they were, and the trade took hold as the principle industry of the area. In 1573, Wilhelm de Wiersberg was the first armorer to be honored with the title of Lord Mayor of the City of Solingen.

Such renowned names as Köller, Grosschelde, Wundes, Lobach and Kirschbaum were but a few of the masters to execute the art over the next hundred years. They remained with the established, but pains-taking techniques of producing their blades by hand. Two centuries of wars caused the production techniques to be improved to meet the increased demand. Less stringent restrictions were imposed on the craftsmen, and many of the artisans saved their hand-forging for only the best quality blades, and turned to the mechanical aids which were born of the Industrial Revolution. The technique of rolling sword blades was introduced in 1847, and is still in use today. This allowed for faster production, but also marked a departure from the hand-forging techniques employed by the swordsmiths.

Industrialization caused the individual craftsman to relocate from the river side of the city, and in many cases, merge their efforts to meet the ever increasing demand. World trade made it necessary to be in close proximity to the rail lines which were to haul the wares to the various trade centers. Out of the smoke of production in Germany, however, loomed the clouds of war. World War I placed a demand on the blade producers unlike anything they had ever experienced. Further refinements in mass production techniques were necessary, and from the need, a process of pre-tempering and rolling the steel into sheets was developed. Heavy drop forges stamped out multiple blades which were then ground, finished and assembled into fighting implements.

With the loss of the war in 1918, military blade production ceased, and the less skilled laborers taken on in the various blade firms were discharged. Economic depression overtook all of Germany followed closely by monetary inflation which left the working class destitute, and the factory owners faced with closed doors.

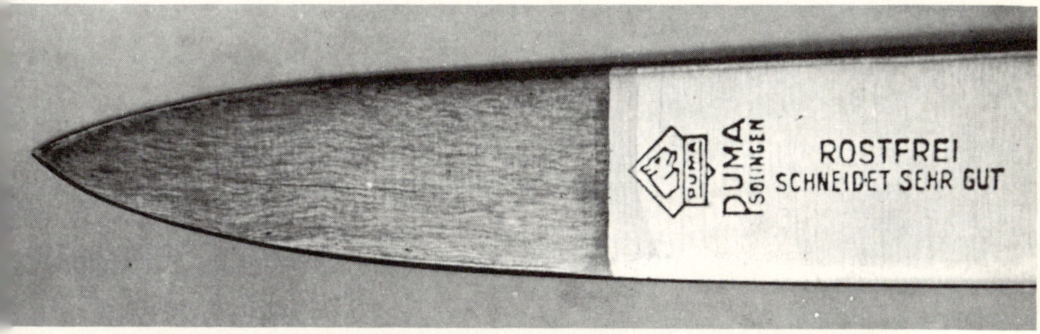

THE PUMA FIRM USED THIS ADVERTISING EXAMPLE TO DEMONSTRATE THE VALUE OF THE STAINLESS STEEL PROCESS WHICH WAS ADOPTED BY THEM AND OTHER MANUFACTURERS IN SOLINGEN. WHEN THE NEWLY TREATED STAINLESS PORTION WAS SHOWN IN CONTRAST WITH THE OLDER POLISHED STEEL PROCESS, IT IS NO WONDER THAT THE CUTLERY BUSINESS RECEIVED A MUCH NEEDED ECONOMIC SHOT-IN-THE-ARM.

The system of guilds (specialized organs which grouped the artisans by trade, and the forerunner of today's system of unions) which existed over the centuries prior to the turn of the twentieth century gave way to meet the demands of mass production. Though some semblance of the guild system remained, the craftsmen surrendered a great deal of their art-form for the sake of mass production. Mass production also meant fewer artisans were required to meet production needs, thus the base of skilled artisans declined.

In order to survive economically, firms stressed the cutlery aspect of their trade. Competition was keen among the many blade producers and each sought to devise new and more appealing products to lure prospective customers to their wares. Out of this competition came stainless steel, enabling the manufacturer to produce tableware which resembled the bright silver finish and eliminated the threat of rust. Foreign demand for presentation swords partially served to hold some of the craftsmen close to their forges.

Hitler's rise to power was largely due to the Weimar government's inability to stabilize the German economy. Hitler appealed to the working class with promises of work and stable economy. When elected as Chancellor of Germany in 1933, he and his Minister of Finance immediately set about to deliver on those promises. Many of the programs instituted by the Weimar Republic were just beginning to take root, and Hitler was quick to take credit for their success. The new Nazi government attempted to infuse new life-blood into the economic veins of the nation. One of his first acts of international tones was the seizure of the Saar region which had been placed under the control of the French government as a result of the Versailles Treaty. The plebicite which returned the Saar to Germany also served to start the steel presses rolling in earnest in the Ruhr and Rhineland industrial regions. Once the restrictions of the Versailles Treaty were no longer considered binding by the Germans, a phantom army grew at an amazing pace and soon became very real. Hitler used the long established tradition of the sword to instill pride in the military machine which had dwindled following a lost war. The blade cartel of Solingen was headed by astute businessmen who sought the profits to be gained from the growning ranks of Army officers. New designs were created and approved, and among others, the "Field Marshal" series was placed into production. Thus gave rise to the "Blücher", Lützow", "Prinz Eugen", "Derfflinger" and "Zeiten" patterns for the Army.

Dagger production rapidly supplanted the sword as the principle production item for the military in Solingen. The first sidearm to be produced in quantity was the Hitler Youth Knife, which had been placed into production even before the election of Hitler....possibly as early as 1929. Once in control of the government, however, Hitler approved a plan put forth by the Solingen blade merchants to bolster the Solingen economy. The first approved Nazi dagger, the SA Service Dagger, was produced in mass quantity to meet the needs of almost three million SA men. After 1933, a rapid succession of new Nazi dagger and sword side-arms were designed and produced for wear by virtually every element of German officialdom.

Acceptance of designs meant production contracts to the manufacturers, and new designs were offered at an unending pace. Designs were submitted, and in some cases prototypes produced, for approval right up to the end of World War II. One series of prototypes was produced by the Eickhorn firm to commemorate the "final victory". The Eickhorn firm was noted for their exceptional designs, and was singled out in 1939, and again in 1940, with the award of the Gaudiploma for excellence.

Acceptance of designs for production was no easy matter, even considering Hitler's desire to stimulate blade production and motivate the men of his military and quasi-military elements. The birth of a dagger or sword started with the manufacturer stating his design concept or approving one submitted by a member of his organization. It was not unusual for a design to originate from among the many designs created by students attending the trade school, as design was part of their course of study. When the artist, who was commissioned to transfer the concept to a piece of finished, presentation art, submitted his work for approval, one or a series of designs would be selected as a basis for prototype production. Metal artisans created the full-scale prototype hilt by hand. Some newly designed hilts were sometimes mounted on a finished blade, but more often than not, only the hilt design was submitted for approval since the reputation for blade production by each manufacturer was already established.

All patterns, regardless of their intended use, were submitted to the Reichszeugmeisterei (RZM) in Berlin or Munich for approval. At the same time, a design patent was applied for but was only granted following approval of the design by the RZM to place the pattern into production. A mold was then produced from the prototype hilt, and subsequent hilt production became a tedious hand-work operation. The new design hilt was then mounted to the finished blade, the production piece again inspected by a representative of the RZM, and upon final approval, the new pattern was placed into mass production. Thereafter, only political daggers, swords and bayonets were inspected by members of the RZM at the end of a days production prior to shipping. All other patterns were produced and shipped without benefit of further examination by the RZM.

Mass orders for official military swords... the Air Force "Fliegerschwert", the Navy "Säbel", and the Army basic "Säbel" without design motif... came from each respective military ministry of the Oberkommando der Wehrmacht, directly from Berlin. Orders for sidearms of a political nature were ordered through the RZM in Munich. All others were not controlled, and could be purchased by the individual directly from the retailer. Selection of the desired design was a matter of choice for the purchaser. Only persons of officer rank, however, were given this freedom of choice. Sword-bearing NCOs were required to purchase and carry only the officially accepted sword for his respective military service, even though exceptions to this policy have been frequently noted.

THE COMMANDER OF THE FLA-ERS-BTL (mot.) 66 TALKS WITH OBERST HAUSCHULZ DURING THE COURSE OF AN INSPECTION. NOTE THAT THE HAUPTMANN CONTINUES TO WEAR HIS WEIMAR-PATTERN SABER WITH STATE INSIGNE AS LATE AS 1941, WHILE THE OBERST WEARS THE STANDARD-PATTERN, ARMY OFFICER' SABER.

From 1935 to the outbreak of war in Europe in 1939, the German military and quasi-military forces grew in considerable strength. This was largely due to the nationalistic nature of the recruitment program and the appeal of the newly designed uniform trappings. The blade sidearm played a small, but significant roll in the overall psychological appeal to the men of military age. Professional military men who were forced to disguise their continued military service by serving in the police or border guard forces simply changed their uniform and surfaced openly after the restrictions of the Versailles Treaty were cast off by Hitler. Some old guard officers continued to carry their Imperial period sword even after returning to formal military service under the Nazis. Others retained some degree of the old traditions by mounting the blade with a new design hilt bearing the swastika motif. Most, however, simply set their old sword aside and purchased one of the many designed swords which appealed to their particular taste. The Solingen manufacturers saw to it that there was an ample selection from which to choose.

The clouds of war which hung over Europe were further darkened by the smoke of the Ruhr steel mills producing the machines of war. In Solingen, bayonets were produced by the tens of thousands alongside the dress sidearms. A year after the outbreak of war, it became apparent that Germany did not have a sufficient stockpile of critical material to sustain war production, and adjustments had to be made in non-essential production to meet the war needs. Such materal as brass, bronze and nickle were declared strategic materials. Thus, in 1940, blade manufacturers were required to produce hilt mountings from aluminum or suitable metal alloys. The prototype designs which required strategic material for their initial manufacture were handmade by a central agency in Berlin. The Berlin facility was granted an exemption from the restriction of the use of strategic materials, and design production continued to the war's end. Prototypes produced by this agency continued to bear the trademark of the firm which had commissioned the undertaking. Even though designs were continually produced, wear of the blade sidearm gave way to the pistol sidearm in 1944, when Germany began her struggle for survival.

The steel producing area of the Ruhr came under the onslaught of saturation bombing by the Allied air forces. The major city of Cologne, some 75 kilometers from Solingen, suffered 90% destruction. Many of the blade firms in Solingen shared the fate, and were wholly or partially destroyed in the Allied effort to eliminate Germany's production capability. Those firms which continued to produce critical implements of war fell into the hands of advancing U. S. troops, and were systematically sacked. At the war's end, Polish military forces were given responsibility for the occupation of Solingen, which remained in effect until 1951. During this period, the Polish soldiers, in search of war souvenirs, were attracted to the beauty of the blade sidearms which were carried in stock storage as a result of the 1944 order forbidding further wear. What the earlier American forces did not remove from the stockpiles, the Poles did. Only small quantities of unassembled parts remained in some of the cellars and attics of the blade firms.

THE KARL EICKHORN FIRM TODAY.

THE PUMA-WERK TODAY.

THE TIGER FIRM TODAY.

HITLER IN HAMBURG, FEBRUARY 14, 1939.

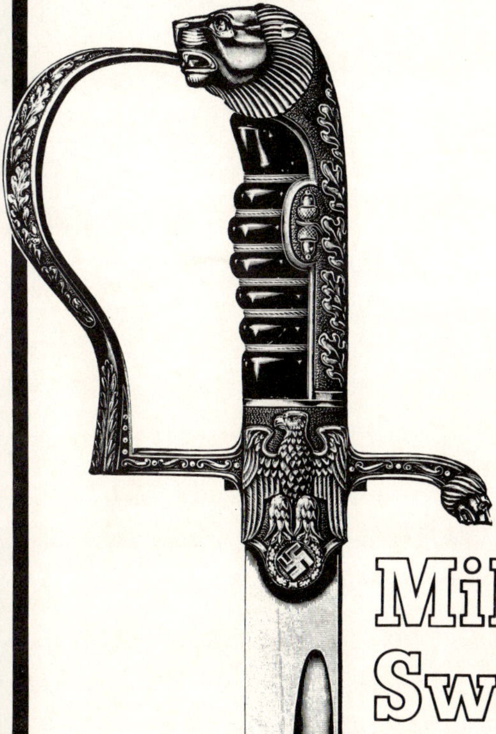

Military Swords

THE ARMY

During the Hitler era, the Army sword reflected the long traditions of by-gone days when the sword or saber sidearm served a functional purpose in combat. Gone were the days when the charging cavalry bore down on their enemy with the sun gleaming on the unsheathed saber blade. Gone also were the days when a family or regimental tradition was reflected in the crest of insigne of the grip. The ornate basket guard which displayed the military unit of the wearer gave way to the less ornate and considerably smaller knuckle-bow. While the tradition of the sword remained with the growing German Army, it reflected the total subjugation of the Army to the absolute control of Adolf Hitler.

While the bayonet continued to serve as a functional combat weapon, the saber was largely relegated to a symbol of one's position. . . . traditionally worn during garrison duty or mounted on the saddle when the mounted parade was performed. No longer the slash and thrust weapon of previous wars, the saber was considerably reduced in weight and size. By the very virtue of its new construction, it lost most of the factors which made it the respected combat weapon.

The Enlisted Man's and Officer's "Säbel" were the only two sabers officially designated by the Army. Both were void of any exterior design motif, and differed basically in size and weight. Officers were authorized, however, to select from among the seemingly limitless, unofficial pattern designs. For this reason, no one branch of military service had a wider variety of sword patterns than did the Army.

The Enlisted Man's "Säbel" and a relatively small quantity of the Officer's "Säbel" were purchased through official channels and stored in the armory for issue by the artificer. Most officers, however, selected to purchase the more ornate unofficial "Säbel" through private channels. For the officer, it simply meant a journey to the retail outlet of his choice, and select from among the many patterns which were placed at his disposal. The variance, therefore, would lie with the personal taste of the individual officer. A general officer might select one of simple design, but which met his needs, while a newly appointed company grade officer might choose from among the fanciest of designs.

The numerous blade firms were highly competitive in creating new design patterns in an effort to attract a larger portion of the consumer market. They offered a wide variety of options to further enhance the attractiveness of the saber and add to the final retail cost. The limitation on the extent of quality of the saber purchased most often depended on the financial status of the purchaser.

Unlike the Luftwaffe, there was no specifically designed saber or sword to distinguish general officer rank. A series of swords was designed, however, for the general officer rank and the rank of Field Marshal in particular, but these pieces never got beyond the design stage before the order to halt blade production was placed into effect.

The regulation governing the wear, etc., of the Officer's "Säbel", cites three different sizes of which the "Säbel" was made available. Selection was limited to the three sizes since the piece was centrally purchased, and an item of general issue. Blade length, however, depended upon the height of the wearer, and was available in lengths from 86cm to 108cm for officer personnel. The Enlisted Man's "Säbel" ranged from 90cm to 112cm.

(Partial translation from "Manual for Ordnance and Supply Personnel", as of April 12, 1940, unrevised printing, Berlin, 1941)

SIDEARMS

Preface:

1. The terms "right, left, front, back, top and bottom" refer to the weapon as it is carried on the left side of the body.
2. The letters and numbers in text refer to Plates 1 and 3. (not shown)
3. The numbers in parenthesis in the text refer to manual identification numbers.

General:

Sidearms are close combat weapons.

Sidearms include the:
Bayonet 84/98 (S. 84/98)
Enlisted Man's Saber (M. S.)
Officer's Saber (O. S.)(1)

[1]The Infantry, Cavalry and Artillery Officer's Saber, and the like found among the troops are for a distinct use, and for that reason are not covered in the regulation.

The saber is used by mounted and dismounted combatants as a cut and thrust weapon.

Each sidearm consists of a blade with handle and a sheath.

The blade consists of:

actual blade with hilt, back, cutting edge, point, and grip section or tang.

The cutting edge...with the exception of the bayonet...will only be sharpened at the time of mobilization ("A. H. M. 37", Sec. 8, Para. 164).

The handle provides for sure manipulation of the weapon. On the M. S. and O. S. it is equipped with a bow guard.

The sheath protects man and horse against injury from the blade, and the blade itself from damage and the effects of weather.

Enlisted Man's Saber:

The blade is made up of the actual blade and the threaded tang.

The actual blade (polished, tempered steel) has a slight rearward curve, moving downward into a double-edged point. The back is flat. On both sides of the blade below the hilt, a groove and cutting edge bevel are milled into it.

The tang is the sharply tapered extension of the blade above the step at the hilt.

Halfway up it is a hole for the pin, and at its upper end are the threads for the tang nut and saber button.

The handle is made up of the:

bow with crossguard
ring
grip
tang nut
cap
pin
saber button

The leather buffer slides off the blade and is not secured.

The bow with crossguard (polished steel) slides over the tang so that the lower hole for the tang with its two projections sits atop the hilt, and the upper hole goes around the tang threads.

The ring (polished steel) encloses the lower end of the grip and the stepped section of the cap. Its lower surface rests on the inside of the crossguard.

The grip (grooved walnut, hardwood, or synthetic wood) slides over the tang, and its stepped end fits into the ring. Its upper end is beak-shaped with a seat for the bow, its center has a hole for the pin and the lower end is stepped to fit into the ring.

The tang nut (steel spanner nut) screws onto the tang threads and holds the bow securely to the grip.

The cap (polished steel) fits on the rear of the grip and protects it from damage. Its upper end, covering the beak-shaped section of the grip, accepts the upper end of the bow, and is fitted with a hole and seat for the saber button. Its lower end slides into the ring with the grip. The two projections are bored for the pin and lie on either side of the grip.

The pin (unpolished steel) is bradded in and joins cap, grip and tang.

The saber button (polished steel spanner nut) screws onto the tang threads over the tang nut, and securely joins the cap to the bow.

The leather buffer (cow leather) is the same size as the top plate of the sheath's mouth-piece. It slides onto the hilt and prevents damage to the crossguard and mount-piece.

The saber sheath consists of the sheath itself with drag and band, mouth-piece with spring strips, and the mouth-piece screws.

The actual sheath (blued drawn seamless tube steel or welded or hard-soldered rolled sheet steel) is shaped to conform to the curve of the blade. The drag (blued steel) is hard-soldered on about 8cm from the upper edge of the sheath.

The mouth-piece consists of the top plate (blued steel), the socket (blued drawn steel tubing) and the spring strips (steel) with rivets (steel).

The socket is hard-soldered to the top plate. The spring strips are attached to the wide sides of the socket with a rivet apiece.

The mouth-piece screws (blued steel with tapered threads) secure the mouth-piece after it has been slipped into the sheath. They are located on the front and rear narrow slides about 4.5mm below the upper rim of the sheath, screwed into the mouth-piece socket.

Officer's Saber:

The Officer's Saber is generally the same in configuration and construction as the Enlisted Man's Saber, but is considerably lighter.

It is made in three different sizes (see size breakout). The size marking 1-3 is stamped on the back of the hilt.

The blade (polished tempered steel) has a rounded back and a proportionately longer double-edged tip.

The tang is not bored for a pin.

The handle is joined to the blade solely by the tang nut, which corresponds to the saber button of the Enlisted Man's Saber. The tang nut is flush with the cap.

The spanner nut (tang nut) which holds on the handle, and the pin which cross-connects the cap, handle and tang on the Enlisted Man's Saber are not present on the Officer's Saber.

The bow with crossguard is made from brass instead of steel.

The finger strap, a two-piece leather loop, slides over the tang between the crossguard and the handle with ring. The forefinger is placed through it when using the weapon.

The ring is made from brass instead of steel.

The grip (grooved white beech) has a black celluloid tube pressed onto it. Braided brass wire is set in the grooves and fastened into a hole at the top and bottom of the grip with wooden pegs. The grip has no hole for a pin.

The saber button is made from brass instead of steel.

The cap is made from brass instead of steel. The projections are not bored for a pin.

The band is only about 7cm from the upper edge of the saber sheath. The sheath has only one threaded hole for a mouth-piece screw located in its front narrow side.

The band has an eyelet with a loose ring instead of a welded-on ring.[2]

The mouth-piece has only one threaded mouth-piece screw hole in its front narrow side.

A mouth-piece screw joins the inserted mouth-piece to the sheath.

Care:

Sidearms may only be disassembled by ordnance personnel at the direction of the responsible ordnance officer or ordnance noncommissioned officer (3)

---

[2] It is interesting to note that the practice of using a secured ring as used during WWI and earlier, was carried over to WWII.

[3] Ordnance Non-Commissioned Officers are those non-commissioned officers who have been ordnance trained at the Army Ordnance School.

The sidearms are to be cleaned after each time they are used. Hand and horse perspiration, moisture, and dirt are to be removed at that time. Blued parts are to be cleaned carefully so that the coating is not damaged. All parts are to be lightly oiled after cleaning.

Wood parts are to be linseed oiled.

Special attention is to be given to drying out the interior of the sheath.

Rust spots and the like are not to be removed from blued or unfinished steel parts. They are only to be repeatedly oiled.

When cleaning the blade, do not put the point against the wall, floor, or the like, in order to keep from bending or blunting, or breaking off the point.

If the temperature changes abruptly, leave the blade in the sheath until the sheath reaches the surrounding temperature. Only then begin cleaning.

Sidearms issued within a unit are to be secured in the wall locker of the person to whom issued in clean condition.

Replacement:

Unuseable sidearm parts or those declared a loss are to be replaced from authorized stockage. Authenticate with a fault slip.

Sidearms declared a loss are to be replaced through service channels.

Size and Weight Data:

M. S.		Length with sheath	36. in.
		Length without sheath	35. in.
		Weight with sheath	3.9 lbs.
		Weight without sheath	2.2 lbs.
O. S.	Size 1	Length with sheath	40.6 in.
		Length without sheath	39.4 in.
	Size 2	Length with sheath	38.2 in.
		Length without sheath	37.0 in.
	Size 3	Length with sheath	34.7 in.
		Length without sheath	33.8 in.

Weight with sheath between 2.6 and 2.3 lbs.
Weight without sheath between 1.5 and 1.4 lbs.

Berlin, April 12, 1940

Army Supreme Command
Army Ordnance Office
Branch for Development & Testing

Koch (signed)

(RIGHT) ALCOSO ARMY OFFICER'S SABER (SABEL), UNATTRIBUTED PATTERN, WITH ETCHED BLADE.

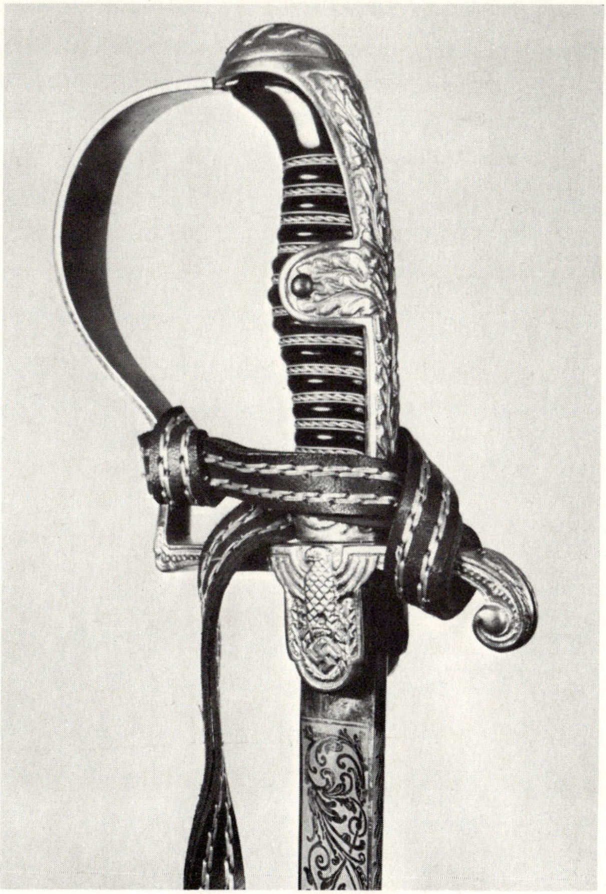

PHOTO: ANDY WALKER

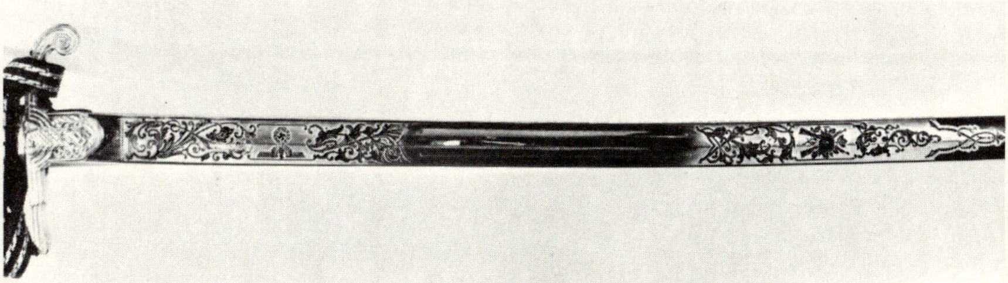

ALCOSO ARMY OFFICER'S SABER (TWO UNATTRIBUTED PATTERNS AND PATTERN NO. 119).

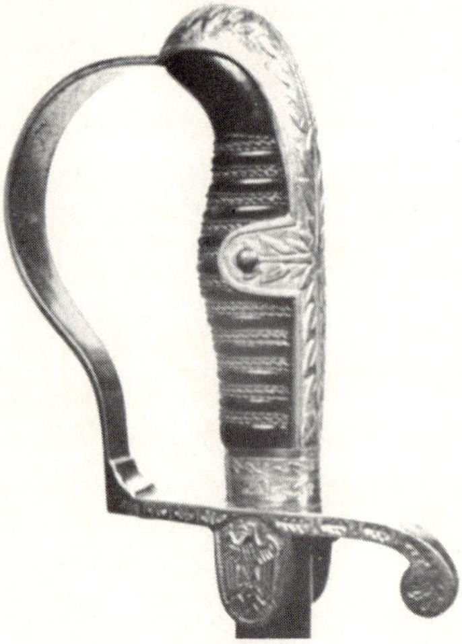

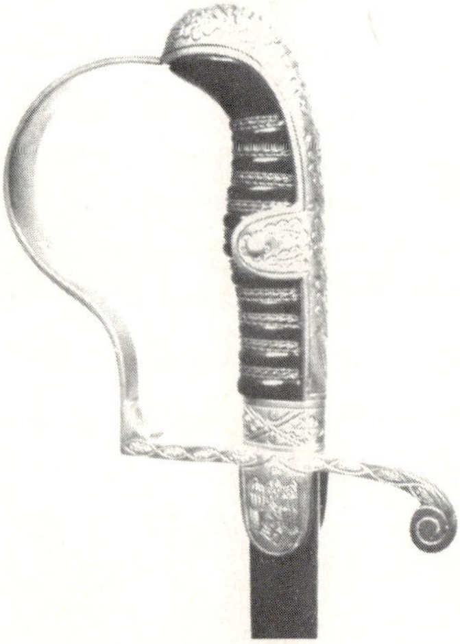

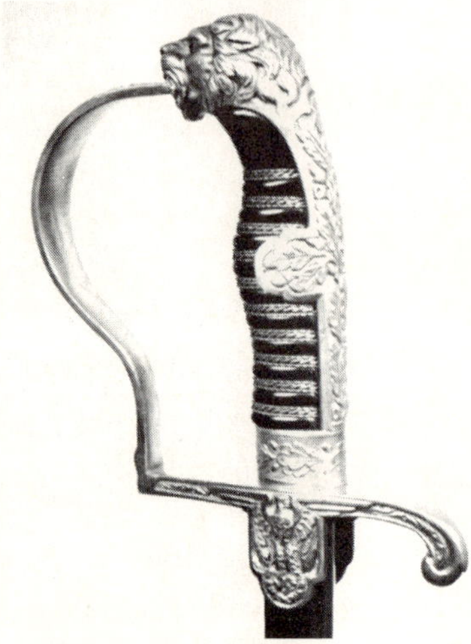

PATTERN NO. 119

PHOTO: ANDY WALKER

ALCOSO ARMY OFFICER'S SABER, UNATTRIBUTED PATTERN, FULLY ENGRAVED HILT.

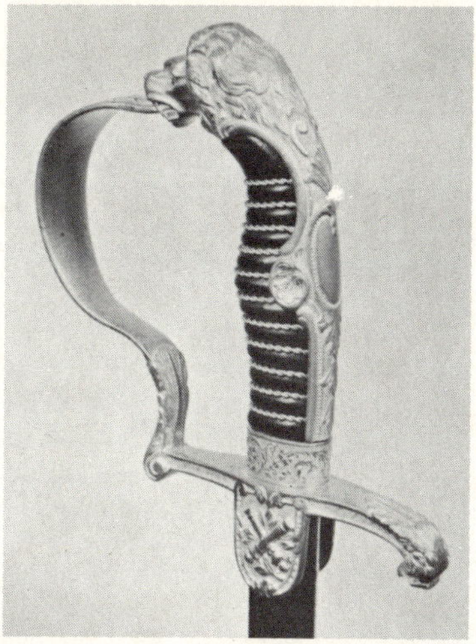

CLEMEN & JUNG ARMY (ARTILLERY) OFFICER'S SWORD, UNATTRIBUTED PATTERN WITH ETCHED BLADE.

A COMPANY COMMANDER, HERE MOUNTED, CARRIED HIS OFFICER'S SABER IN A MOUNTED HARNESS IN A MANNER PRESCRIBED BY REGULATIONS FOR MOUNTED TROOPS.

THIS ARMY OFFICER WAS MARRIED WEARING HIS DRESS UNIFORM TO INCLUDE HIS SABER. NOTE THAT THE OFFICER ELECTED TO USE THE 42cm PORTEPEE NORMALLY WORN WITH THE DAGGER SIDEARM.

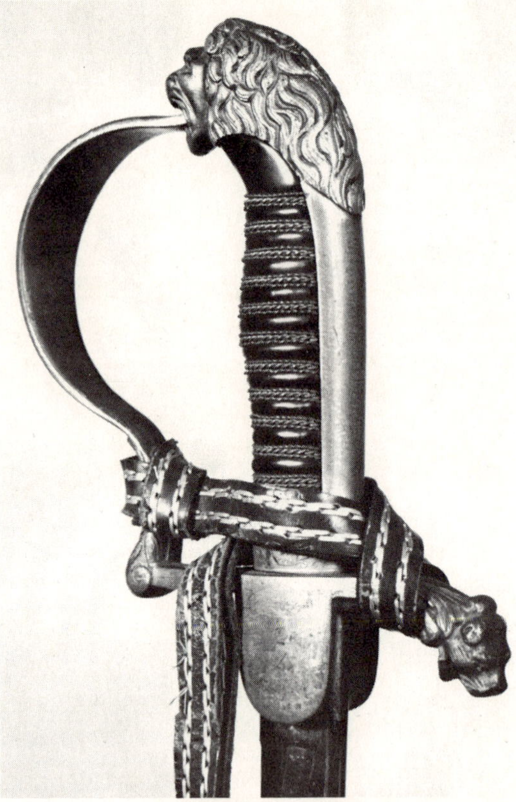

PHOTO: ANDY WALKER

EICKHORN (EARLY PATTERN) ARMY OFFICER'S SABER, UNATTRIBUTED PATTERN, WITH ETCHED BLADE. THE EARLY PATTERN OF THE EICKHORN TRADEMARK WOULD INDICATE THAT THIS SABER WAS MANUFACTURED IN APPROXIMATELY 1930, AND WOULD ACCOUNT FOR THE LACK OF SWASTIKA MOTIF ON THE LANGET.

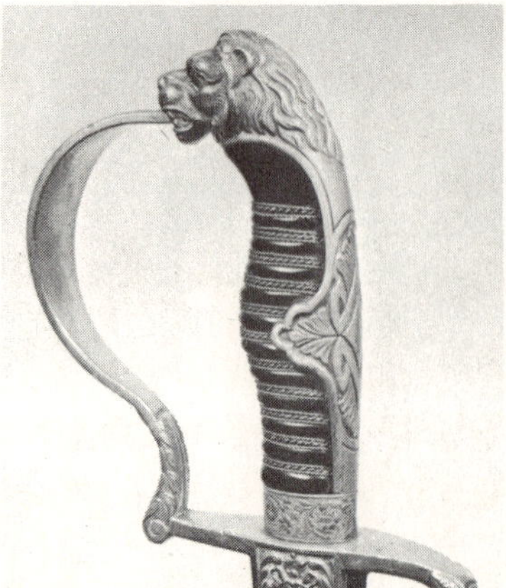

EICKHORN ARMY (CAVALRY) OFFICER'S SABER, NO. 1324.

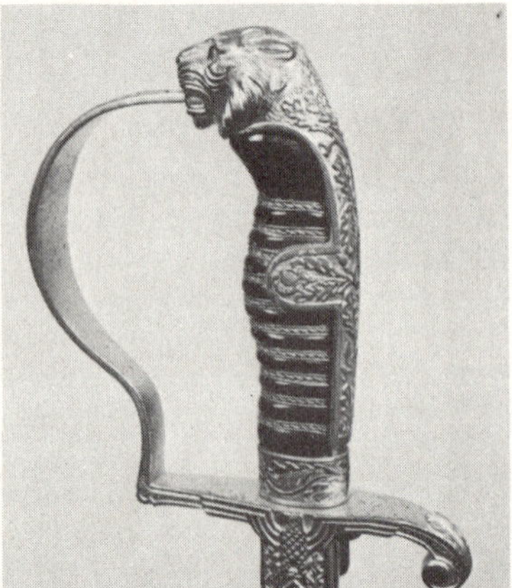

EICKHORN ARMY OFFICER'S SABER, NO. 1695 WITH ETCHED BLADE.

EICKHORN FIELD MARSHAL SERIES

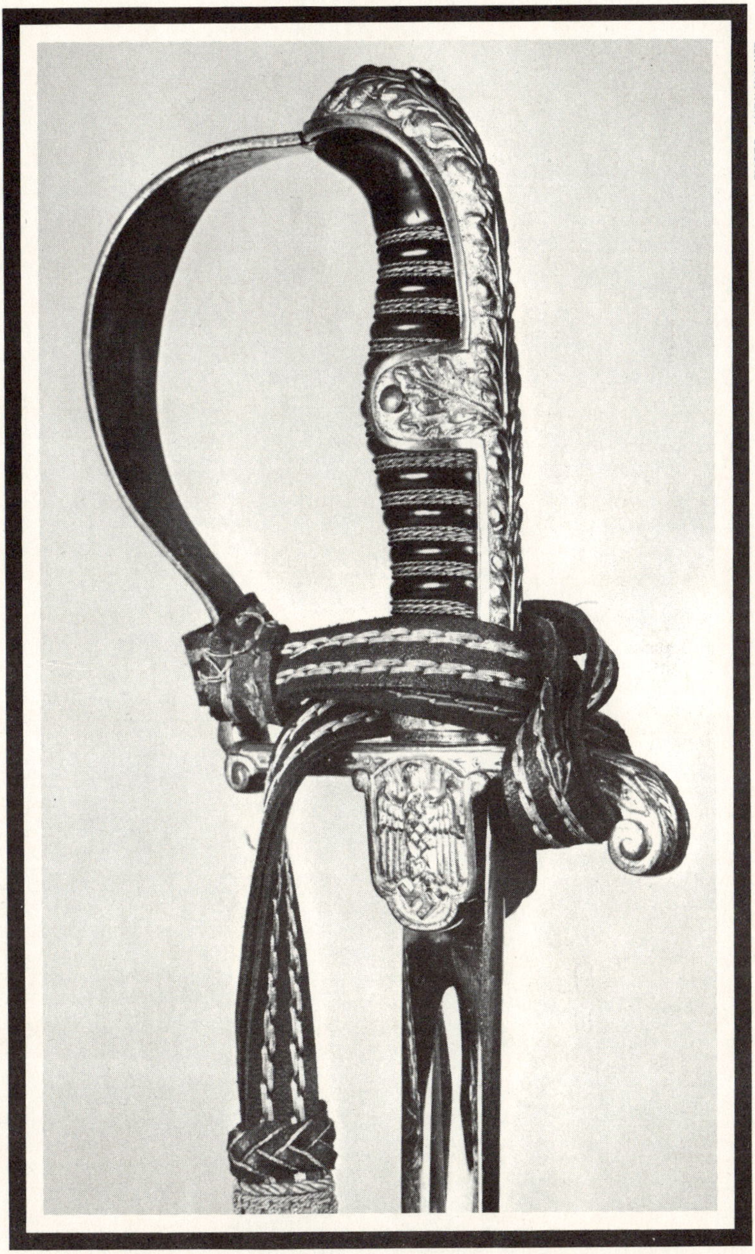

PHOTO: ANDY WALKER

PATTERN NO. 1714 "FREIHERR VON STEIN", WITH ETCHED BLADE.

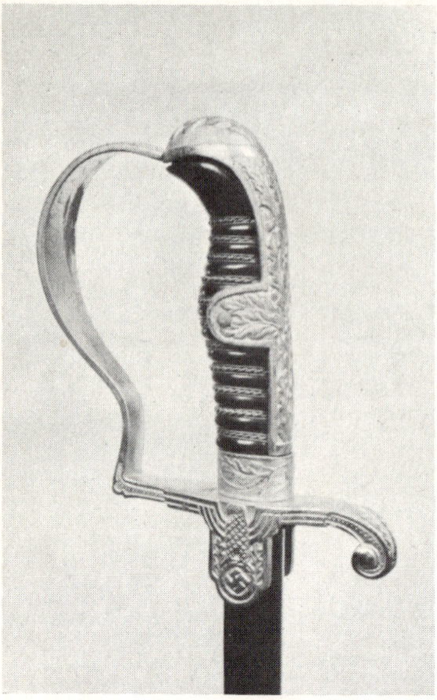

PATTERN NO. 1693 "WRANGEL"

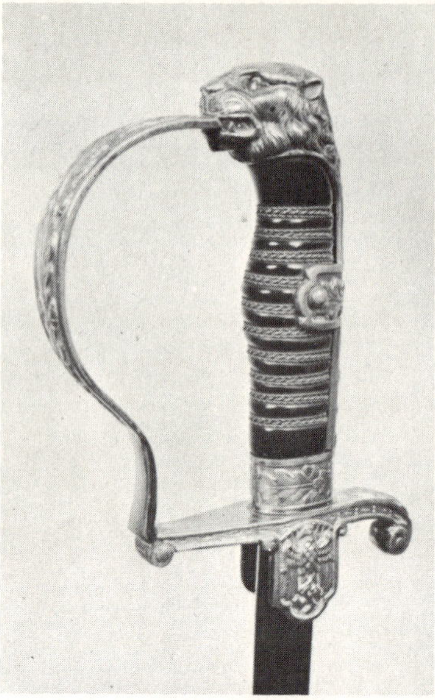

PATTERN NO. 1706 "SCHARNHORST"

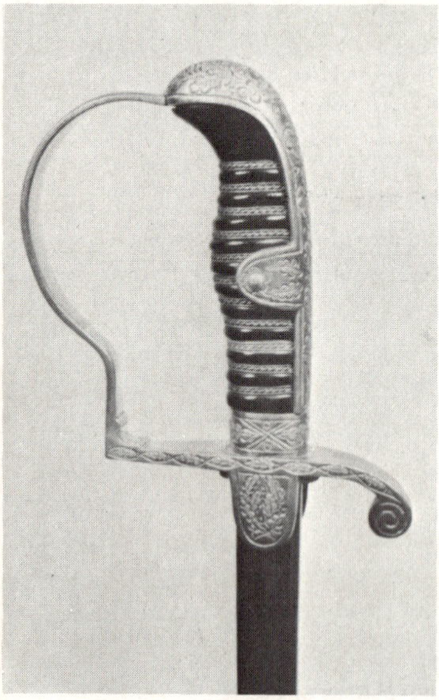

PATTERN NO. 1716 "ROON"

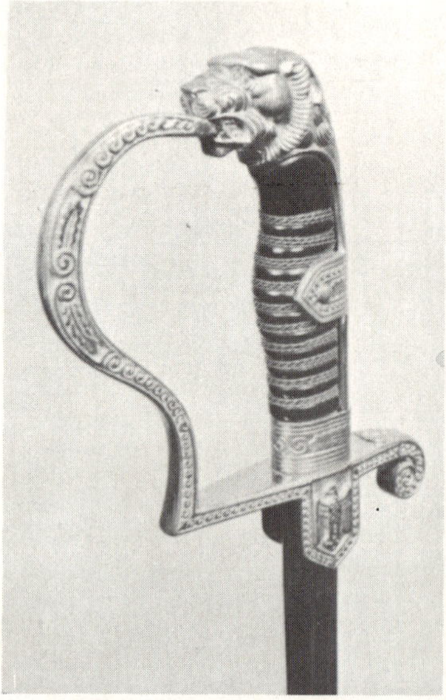

PATTERN NO. 1734 "ZIETEN"

PATTERN NO. 1735 "DERFFLINGER"

THE SWORD KNOT WAS A REGULATION ATTACHMENT FOR THE HILT MOUNTING, AUTHORIZED FOR WEAR BY ALL RANKS FROM UNTEROFFIZIER MIT PORTEPEE TO GENERALFELDMARSCHALL.

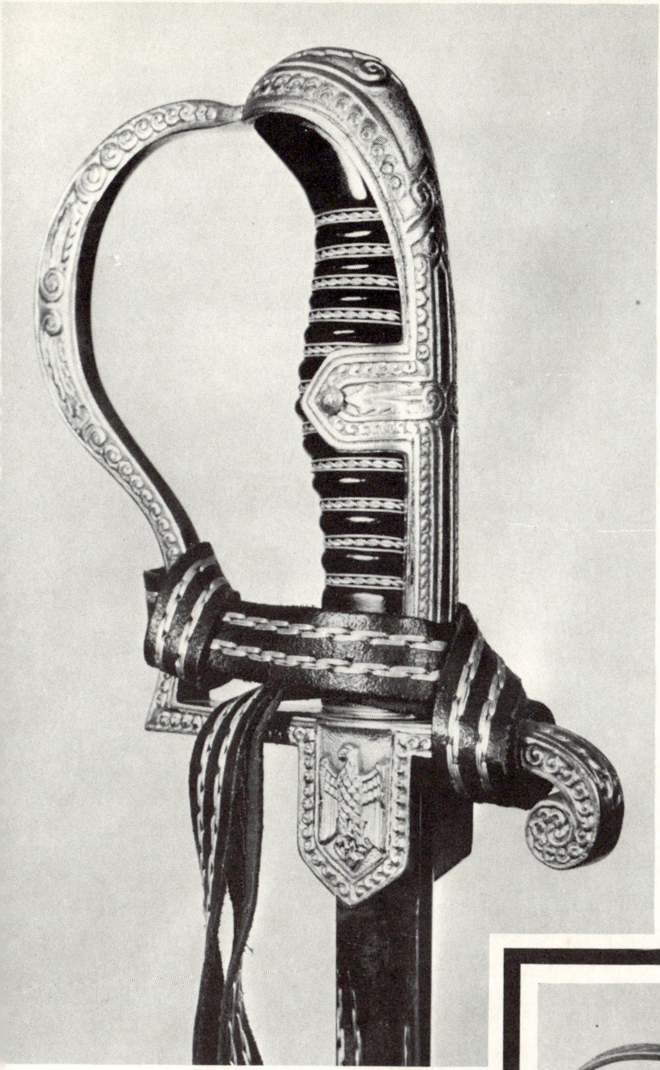

PHOTO: ANDY WALKER

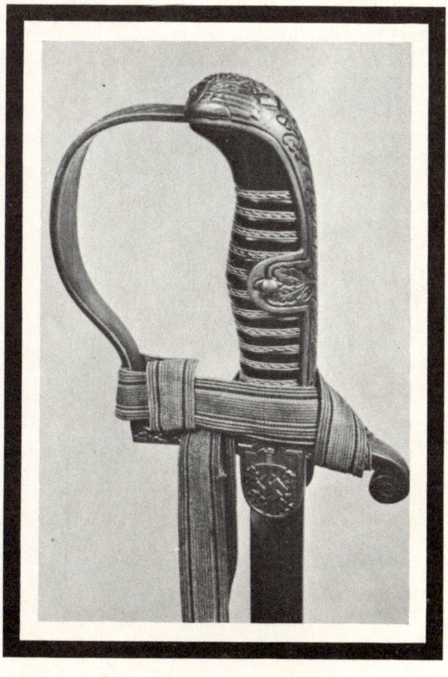

PATTERN NO. 1767 "LÜTZOW", NOTE WEHRMACHT EAGLE DECORATING THE POMMEL. THE KNOT IS A STATE REPRESENTATION (GOLD WITH BROWN STRIPES).

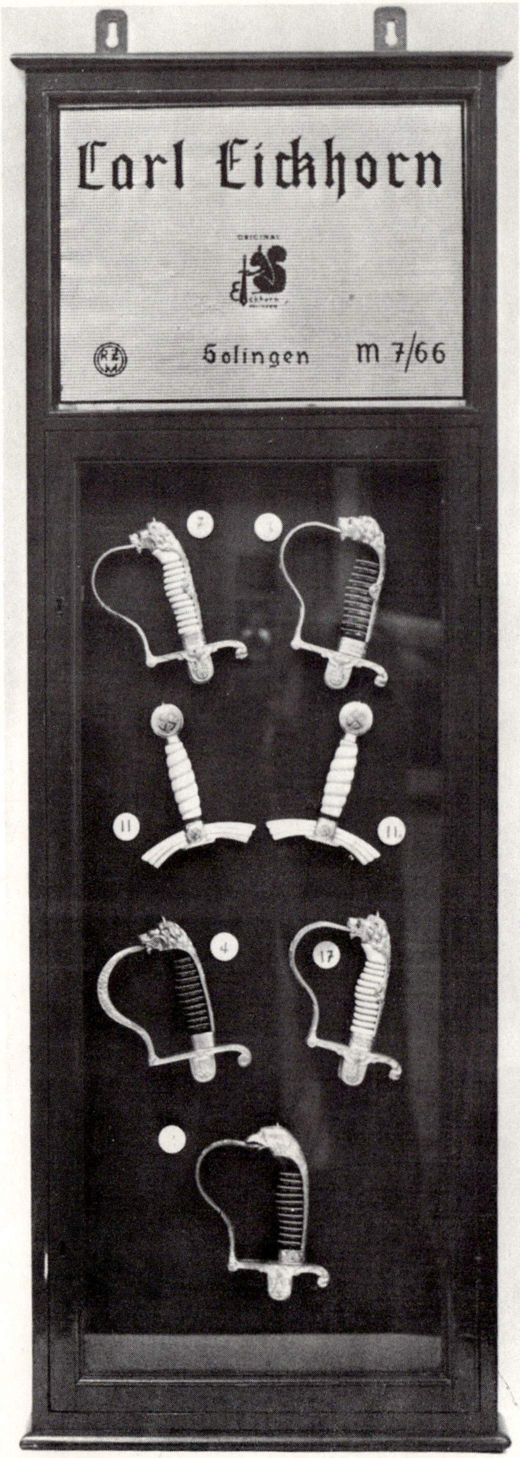

DISPLAY CASE OF PROTOTYPE SWORD HILTS MADE BY THE EICKHORN FIRM FOR PRESENTATION TO THE "RZM" FOR APPROVAL AND EVENTUAL MANUFACTURE. THESE WERE MADE LATE IN THE WAR IN "ANTICIPATION OF THE FINAL VICTORY". THE FINAL VICTORY, HOWEVER, NEVER CAME. NOTE THE WHITE GRIPS ON PATTERNS 7, 11, 11a and 17. THERE IS NO DISCERNIBLE DIFFERENCE BETWEEN PATTERNS 11 AND 11a, AND NO ADDITIONAL INFORMATION WAS AVAILABLE AT THE EICKHORN FIRM TO SOLVE THIS PUZZELING QUESTION.

 PHOTO: ANDY WALKER

RICHARD HERDER ARMY OFFICER'S SWORD, NO. 1017, WITH SILVER EAGLE/SWASTIKA ON GOLD HILT. MODEL NO. 1007 IS THE SAME PATTERN, BUT WITHOUT THE SILVER EAGLE/SWASTIKA (i.e. THE ENTIRE HILT MOUNT IS GOLD). FURTHER, IT IS INTERESTING TO NOTE THAT THIS PATTERN CARRIES THE DESIGNATION OF "DEGEN" RATHER THAN "SABEL", EVEN THOUGH IT HAS THE CURVED BLADE. AFTER 1938, THE MAIN FEATURE OF THE HERDER DESIGN WAS ITS MODERNISTIC FLARE, AND THE USE OF THE SILVER WEHRMACHT EAGLE ON THE GOLD CROSSGUARD. (NOTE: SAME PATTERN, BUT WITHOUT LION HEAD (WITH SILVER EAGLE) DESIGNATED NO. 1019, AND WITHOUT SILVER EAGLE NO. 1009).

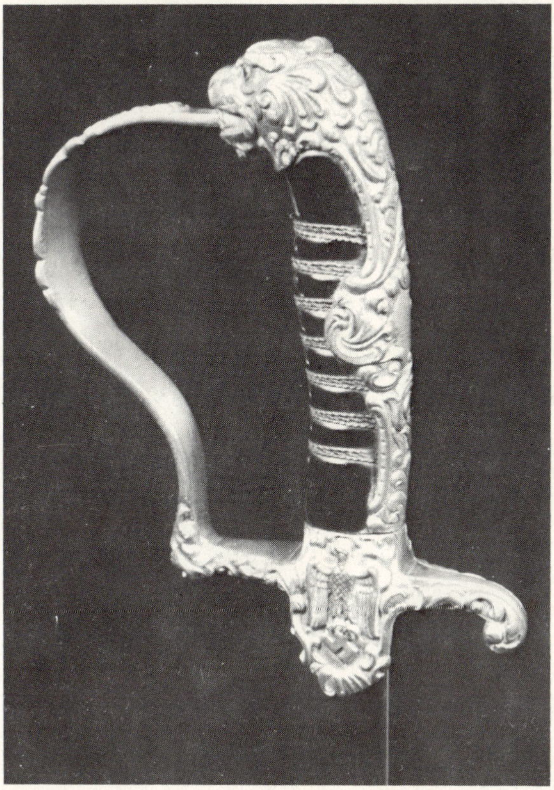

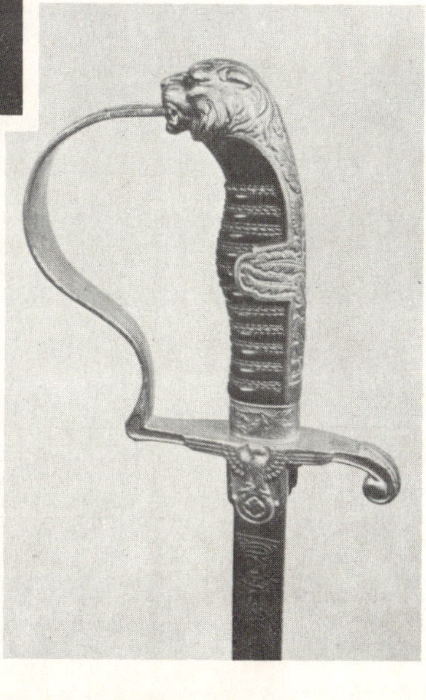

RICHARD HERDER ARMY OFFICER'S SABER, UNATTRIBUTED PATTERN WITH ETCHED HONOR PATTERN BLADE.

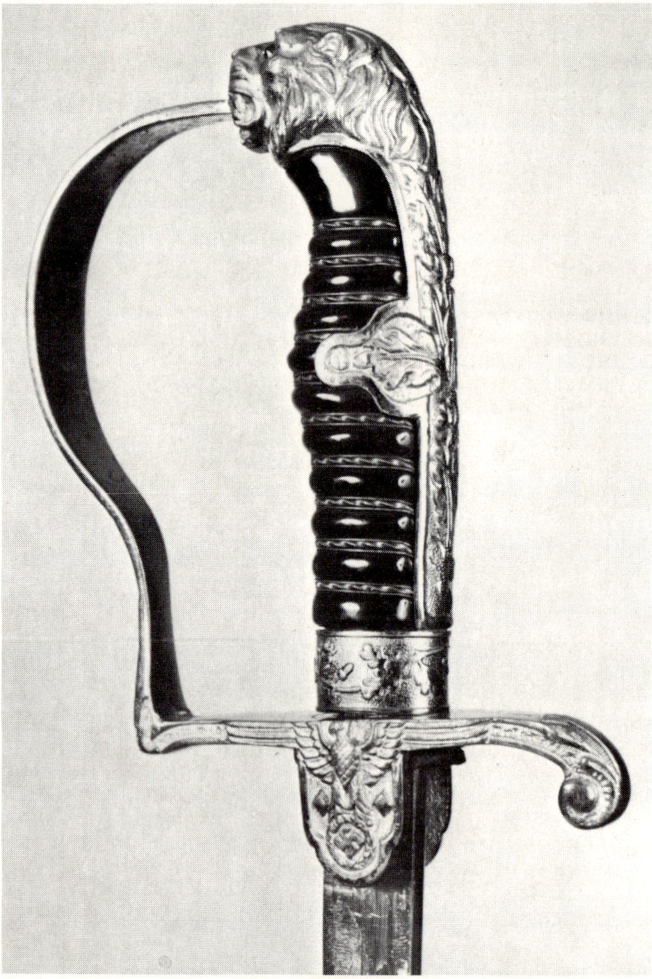

PHOTO: ANDY WALKER

F.W. HÖLLER ARMY OFFICER'S SABER, UNATTRIBUTED PATTERN.

CAPT. R. J. COLE COLLECTION

F.W. HÖLLER ARMY OFFICER'S SABER NO. 60, WITH PRESENTATION BLADE.

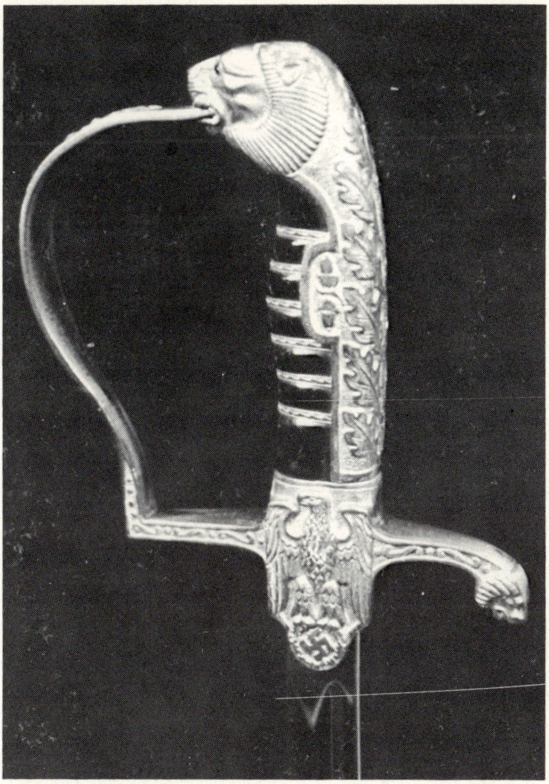

F.W. HÖLLER ARMY OFFICER'S SABER NO. 21.

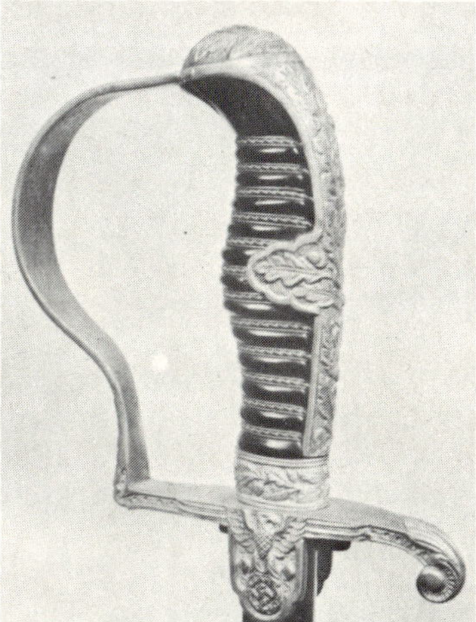

F.W. HÖLLER
ARMY OFFICER'S SABER NO. 24.

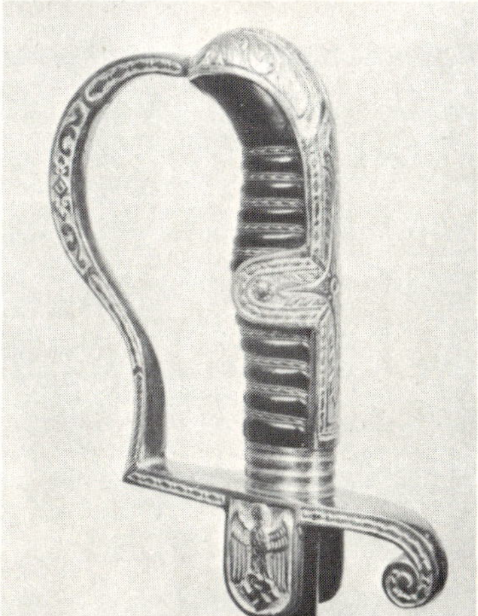

F.W. HÖLLER
ARMY OFFICER'S SABER NO. 61.

GARY BARNUM COLLECTION

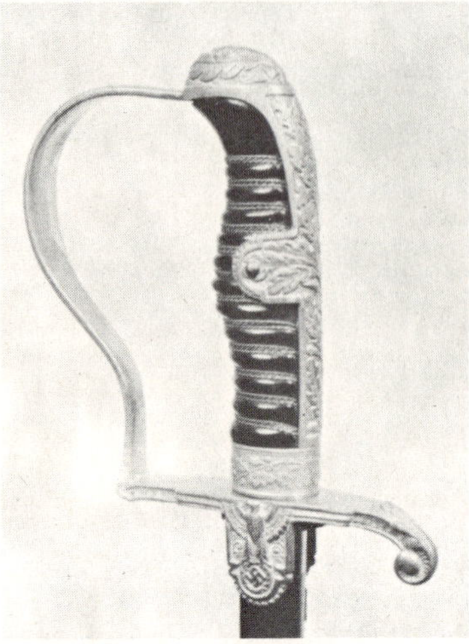

E&F HÖRSTER ARMY OFFICER'S SABER, UNATTRIBUTED PATTERN.

E & F HÖRSTER ARMY OFFICER'S SABER, UNATTRIBUTED PATTERN, WITH ETCHED BLADE. THE ETCHED BLADE WAS AN ITEM OF OPTIONAL PURCHASE AND ADDED CONSIDERABLY TO THE FINAL COST. THE ETCHED PATTERN USUALLY REFLECTED THE SERVICE ARM TO WHICH THE WEARER BELONGED, i.e. INFANTRY, ARTILLERY, ARMOR, ENGINEER, ETC. WHILE MOST BLADES WERE OF A BASIC PATTERN, INDIVIDUAL PATTERNS WERE CREATED FOR SPECIALLY AWARDED "HONOR" PRIZES.

PHOTO: MARC BROWN/HICKOX COLLECTION

KLAAS ARMY OFFICER'S SABER, UNATTRIBUTED PATTERN. THE KLAAS FIRM PLACES THE DATE OF MANUFACTURE AT 1942.

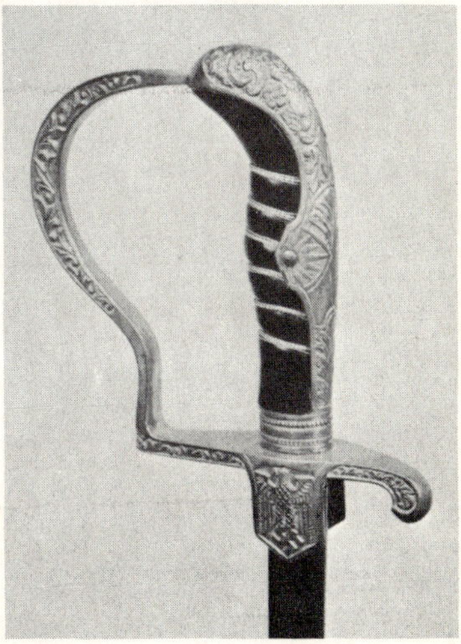

KLAAS ARMY OFFICER'S SABER, UNATTRIBUTED PATTERN.

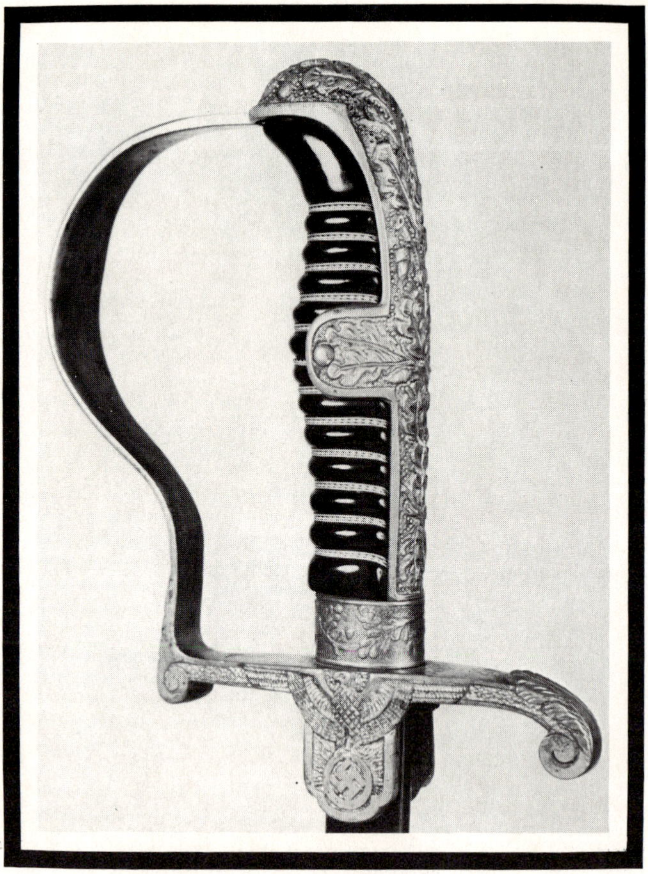

PHOTO: ANDY WALKER

KLAAS ARMY OFFICER'S SABER NO. 233.

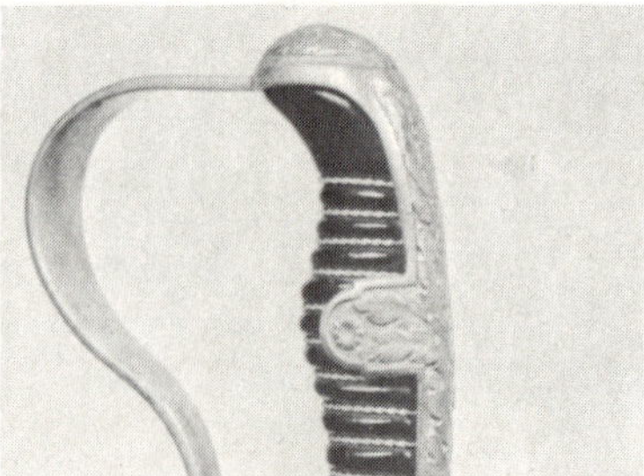

KREBS ARMY OFFICER'S SABER NO. 03.

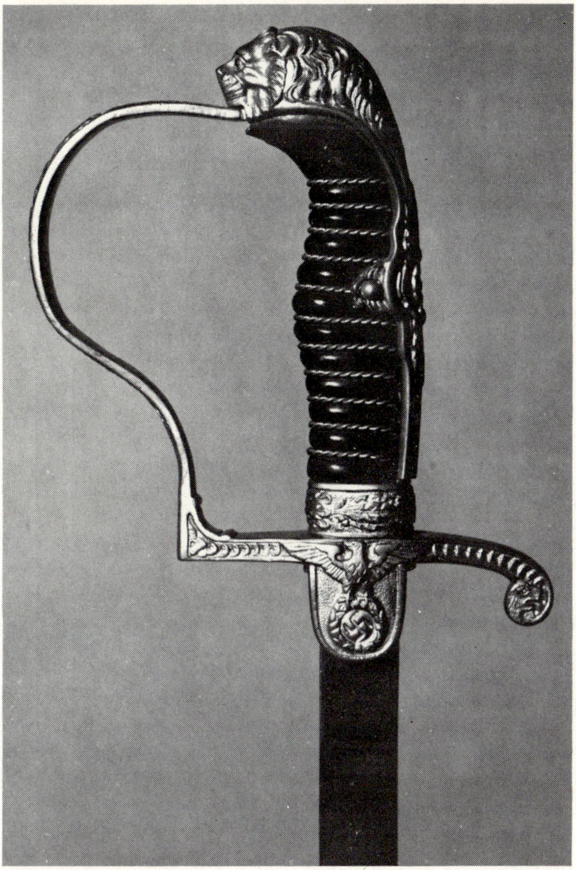

GARY BARNUM COLLECTION

P.D. LÜNESCHLOSS ARMY OFFICER'S SABER, UNATTRIBUTED PATTERN.

E. PACK ARMY OFFICER'S SABER NO. 204.

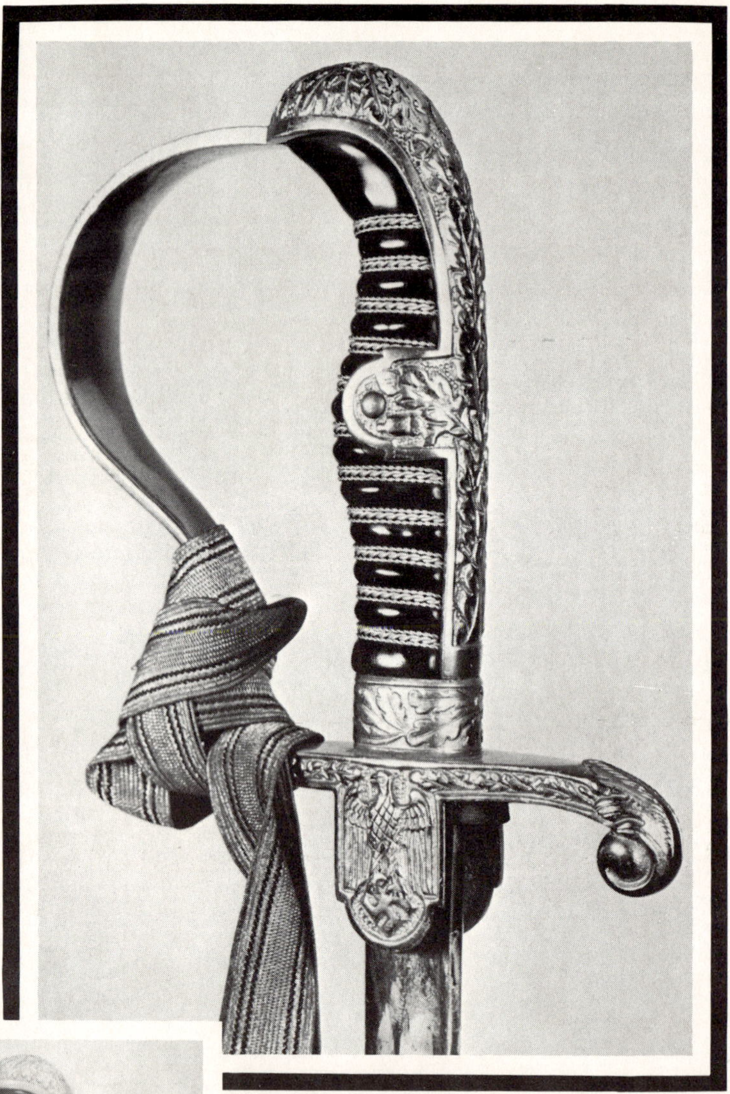

PUMA ARMY OFFICER'S SABER, UNATTRIBUTED PATTERN (KNOT SHOWN IS A STATE REPRESENTATIVE PIECE WITH GOLD STRAPS AND BROWN STRIPES).

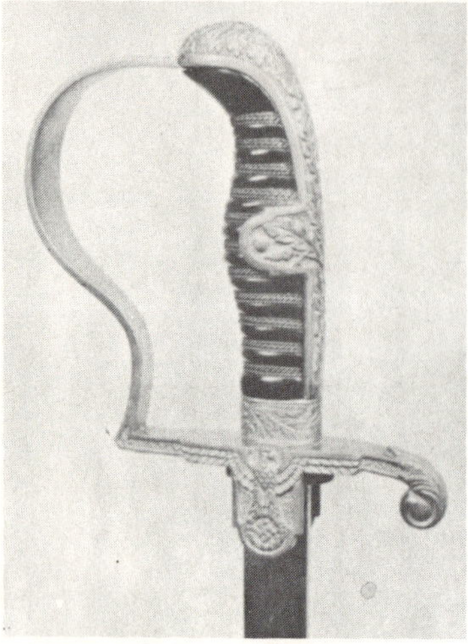

PUMA ARMY OFFICER'S SABER, UNATTRIBUTED PATTERN.

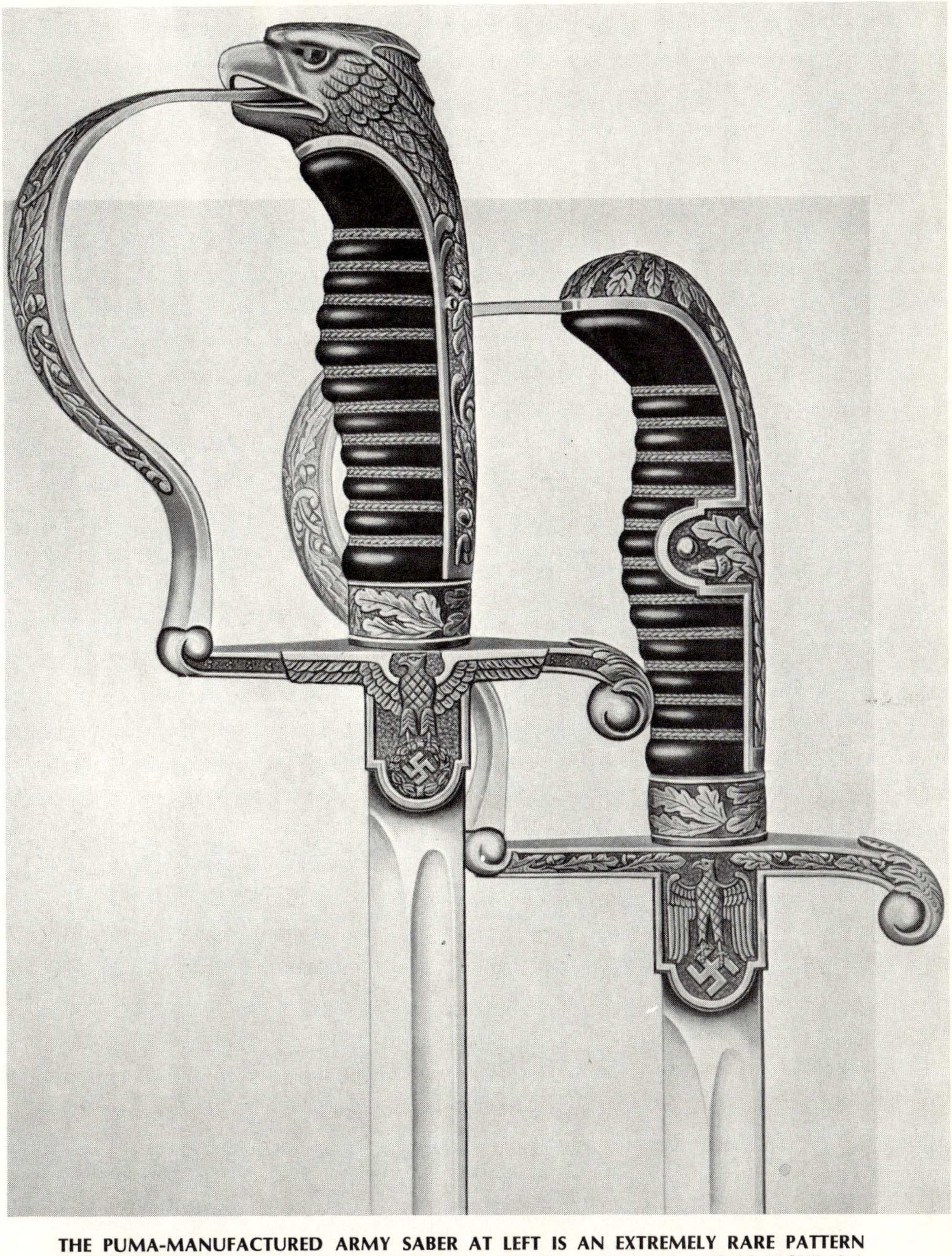

THE PUMA-MANUFACTURED ARMY SABER AT LEFT IS AN EXTREMELY RARE PATTERN DESIGNATED FOR ADMINISTRATIVE OFFICIALS (NO. 30), WHILE THE ONE AT RIGHT IS A STANDARD ARMY OFFICER'S SABER NO. 33. (ORIGINAL PHOTO COMPLIMENTS OF PUMA-WERK).

PAUL SEILHEIMER ARMY OFFICER'S SABER NO. 85, WITH STRAIGHT ETCHED BLADE. THE BLADE FOUND IN THIS SPECIMEN IS A REPLACEMENT BLADE WHICH STEMS FROM THE IMPERIAL PERIOD, AND REFLECTS HONORABLE PRIOR SERVICE BY A FAMILY MEMBER, IF NOT THE WEARER HIMSELF.

GARY BARNUM COLLECTION

EMIL VOSS ARMY OFFICER'S SABER, UNATTRIBUTED PATTERN (WITH WEHRMACHT EAGLE ON KNUCKLE BOW). GRIP IS BROWN WOOD RATHER THAN THE NORMAL BLACK CELLULOID.

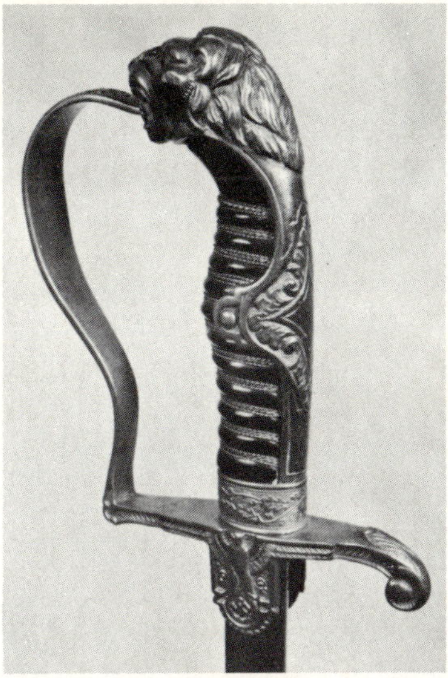

WEYERSBERG ARMY OFFICER'S SABER, UNATTRIBUTED PATTERN.

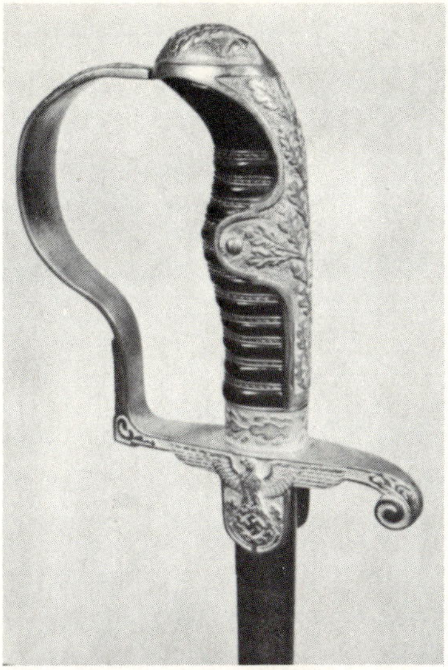

WEYERSBERG ARMY OFFICER'S SABER, UNATTRIBUTED PATTERN.

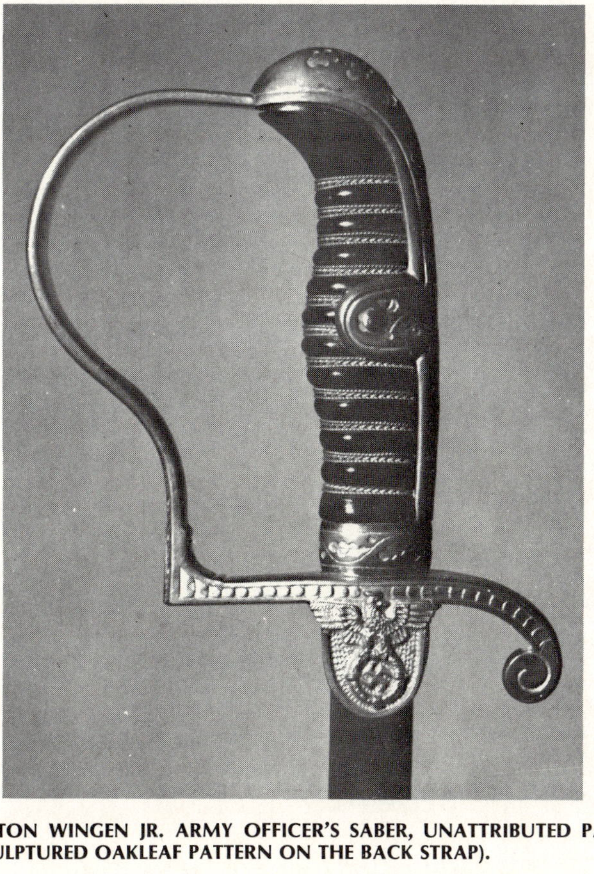

GARY BARNUM COLLECTION

ANTON WINGEN JR. ARMY OFFICER'S SABER, UNATTRIBUTED PATTERN (NOTE DEEPLY SCULPTURED OAKLEAF PATTERN ON THE BACK STRAP).

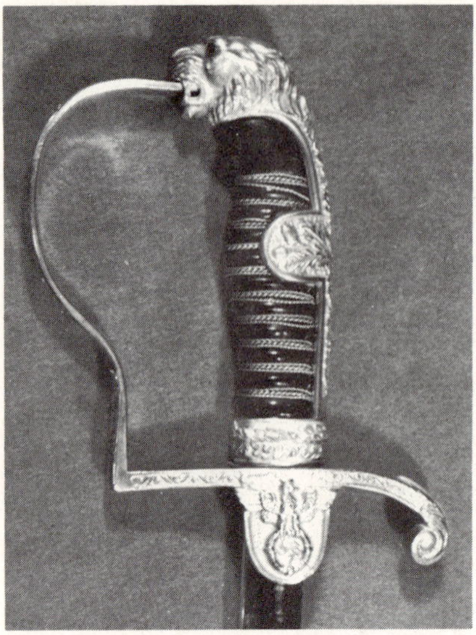

PHOTO: TOM SAUERS

ANTON WINGEN, JR. ARMY OFFICER'S SABER, UNATTRIBUTED PATTERN (SAME AS PRECEDING, ONLY WITH LION HEAD POMMEL).

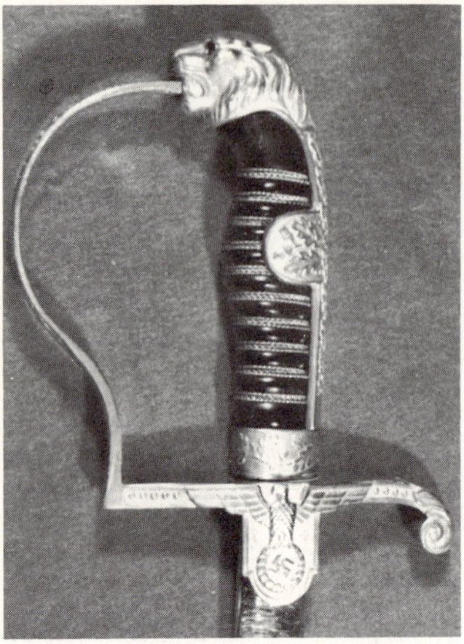

PHOTO: TOM SAUERS

ANTON WINGEN, JR. ARMY OFFICER'S SABER, UNATTRIBUTED PATTERN.

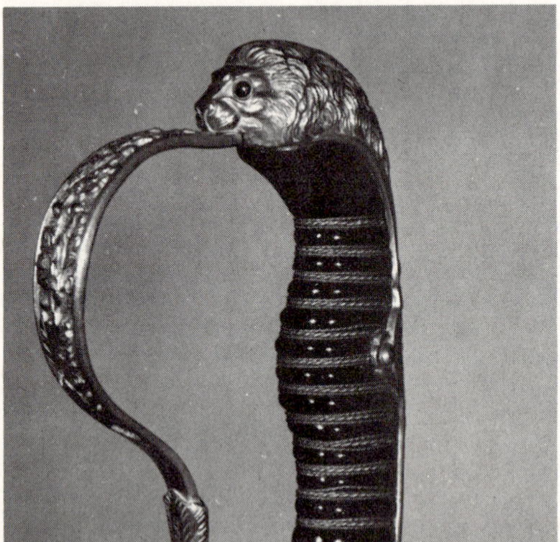

PHOTO: MARC BROWN/HICKOX COLLECTION

WKC UNATTRIBUTED PATTERN ARMY OFFICER'S SABER (NOTE UNUSUAL POMMEL PATTERN).

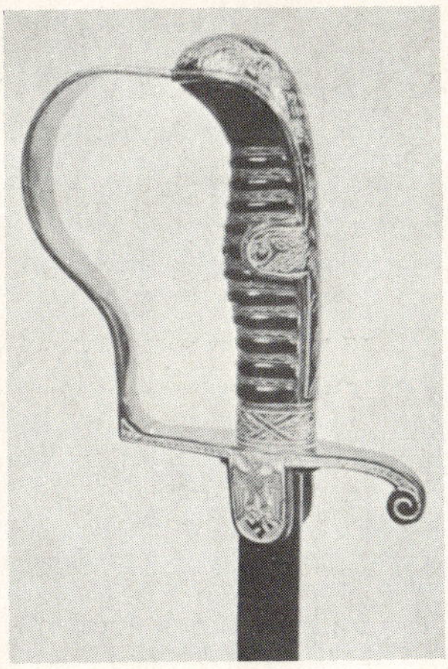

WKC ARMY OFFICER'S SABER NO. 29a, ENGRAVED HILT WITH CAST RICASSO.

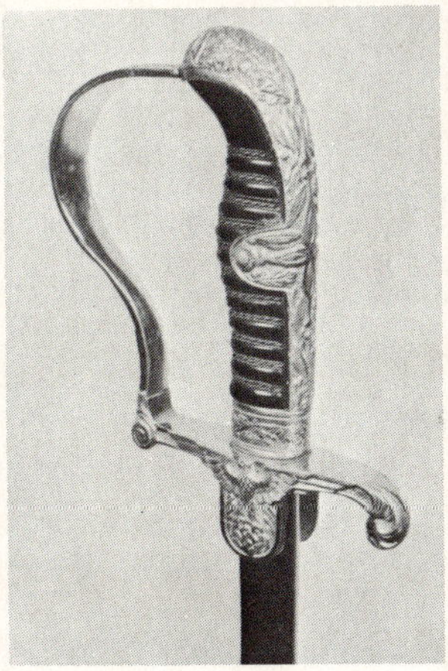

WKC ARMY OFFICER'S SABER NO. 1058.

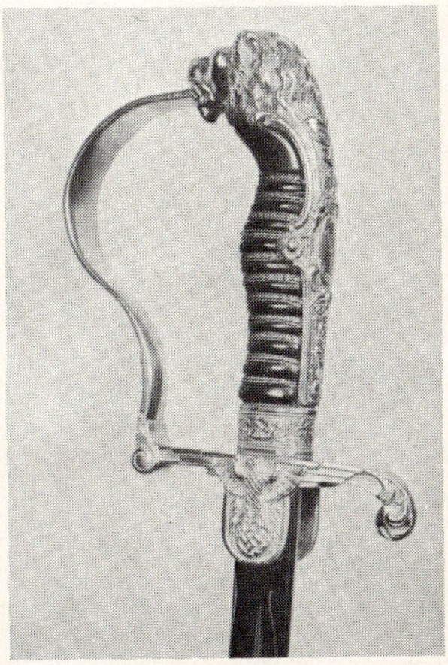

WKC ARMY OFFICER'S SABER NO. 1059.

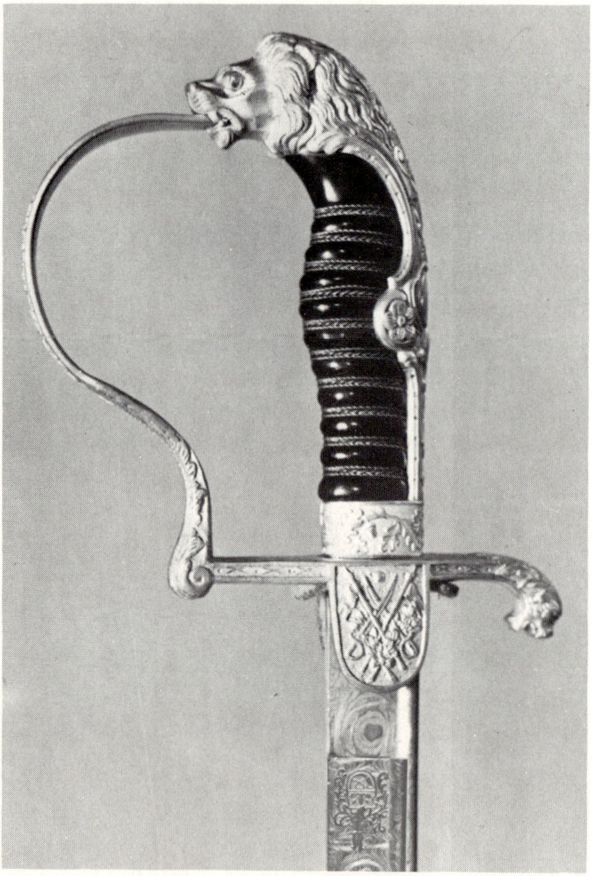

PHOTO: MARC BROWN

UNATTRIBUTED ARMY (CAVALRY) OFFICER'S SABER WITH DAMASCUS BLADE. NOTE VERY ORNATE RAISED CREST IN THE DAMASCUS PATTERN.

PHOTO: ANDY WALKER

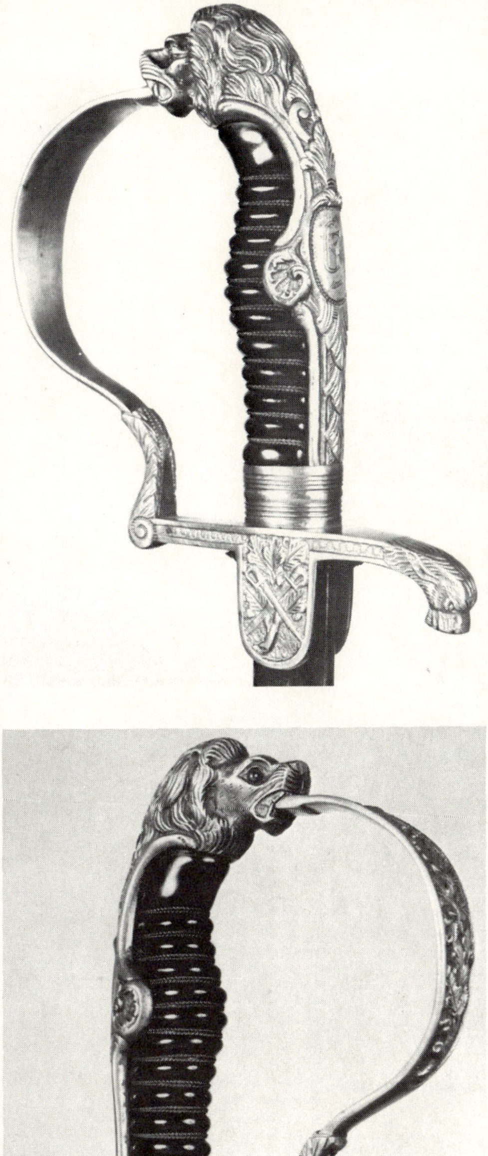

UNATTRIBUTED ARMY (CAVALRY) OFFICER'S SABER (OBVERSE AND REVERSE). NOTE OWNER'S INITIALS ON THE MONOGRAM SHIELD ON THE BACK STRAP.

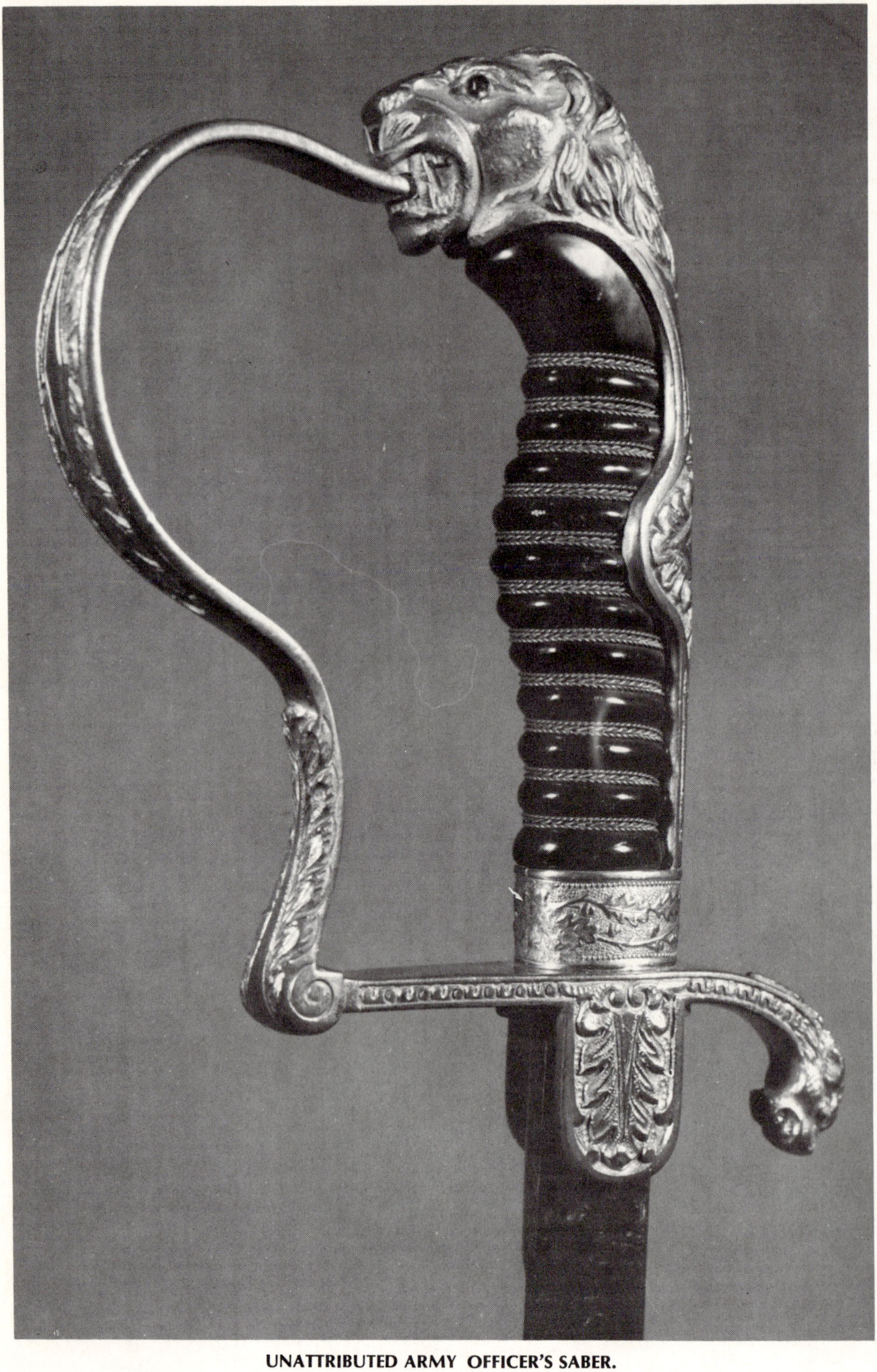

GARY BARNUM COLLECTION

UNATTRIBUTED ARMY OFFICER'S SABER.

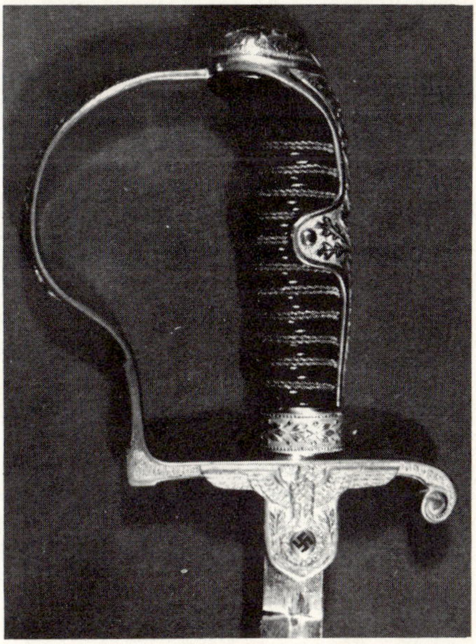

THIS ARMY PATTERN SABER INCORPORATES AN ENAMEL SWASTIKA INTO THE MOTIF OF THE NATIONAL EMBLEM. . .NOT A COMMON PRACTICE SINCE IT WOULD REQUIRE A TWO-STEP PRODUCTION PROCESS FOR THE MANUFACTURE OF THE CROSSGUARD AND KNUCKLE BOW.

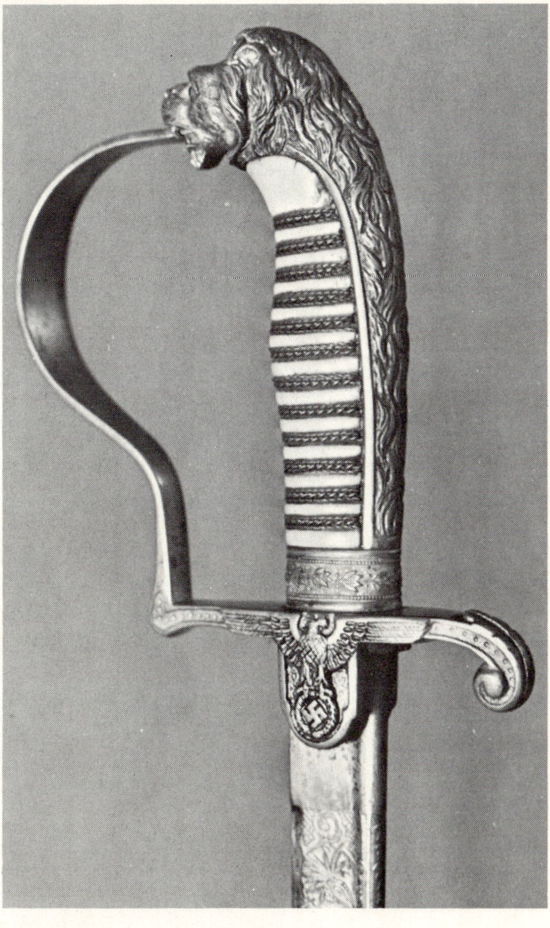

PHOTO: MARC BROWN/HICKOX COLLECTION

UNATTRIBUTED ARMY OFFICER'S SABER, WITH ETCHED PATTERN DAMASCUS BLADE AND WHITE IVORY GRIP. WHILE THE BASIC PATTERN INDICATES THAT THIS IS AN ARMY PATTERN SABER, THE GREEN AND RED EYES IN THE LIONHEAD, THE IVORY GRIP, AND THE ENGRAVED PATTERN WOULD ALSO INDICATE THAT THERE IS CONSIDERABLE NAVY INFLUENCE. COMMENTS ARE WELCOMED.

THE AIR FORCE

On March 5, 1935, Hitler decreed the official establishment of the new German Luftwaffe in blatant contradiction to the limitations and restrictions imposed by the Versailles Treaty. The previous year had seen the introduction of the Luftwaffe "Fliegerschwert", which was designated as the official sword sidearm of the Luftwaffe. It was authorized for wear by senior NCOs (Unteroffizier mit Portepee) and upwards.

The sword resembled those carried by the Teutonic Knights, having a straight, double-edged blade, and a rather simple hilt design. Early patterns had metal fittings of nickel, but those manufactured after 1940 were constructed from aluminum due to the restrictions imposed on strategic materials.

To facilitate purchasing by the central purchasing agency, those swords retained as unit property were ordered through Army channels from 1938 to 1941. Persons qualified to wear the sword, but not authorized to be issued one, were required to purchase their own directly through a retail outlet.

The "Fliegerschwert" was worn by all authorized ranks until late 1935, when a specially designed sword was created and approved for Luftwaffe generals. The general officers' sword had a double-edged, etched blade, ornate brass hilt mountings, and silver wire wrapped handle with a gold-colored Luftwaffe eagle affixed to the front center. This pattern was supplanted with a newly designed generals' "Fliegerdegen" in 1938. Basically the same configuration, the main distinguishing features were the large Luftwaffe eagle in silver on the hilt guard, and the orange solid plastic handle with gold(bronze) wire wrap.

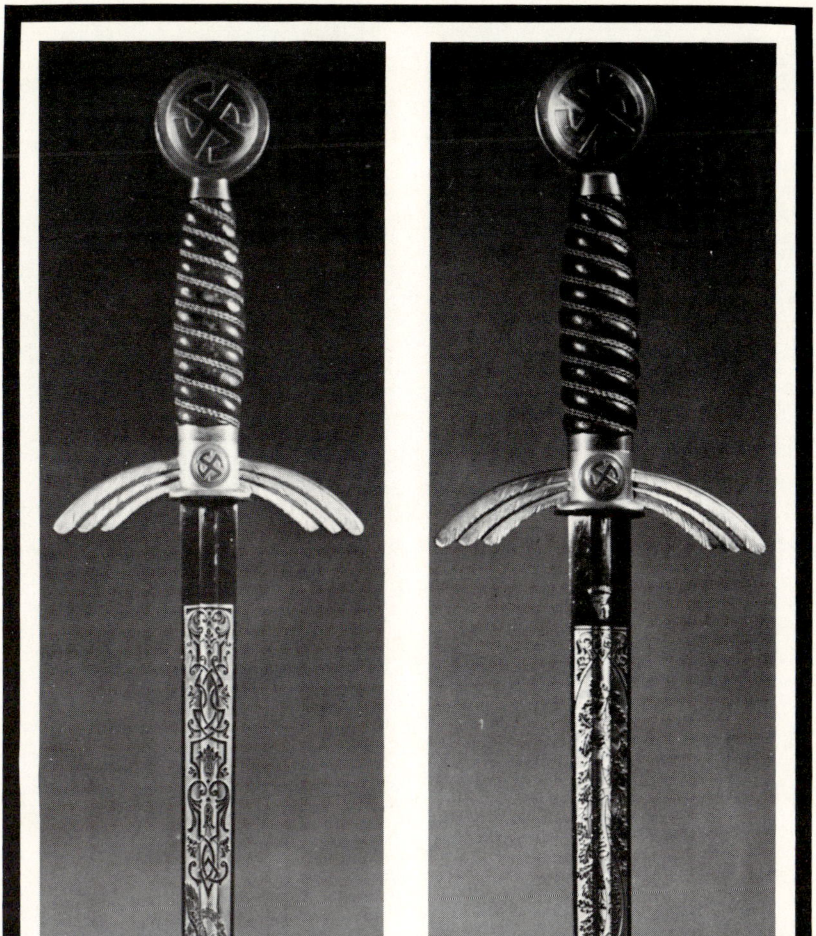

GARY BARNUM COLLECTION

SMF 1934 PATTERN LUFTWAFFE "FLIEGERSCHWERT" WITH DOUBLE-ETCHED BLADE.

As translated from "Anzugordnung für die Luftwaffe", 27.11.35, para. 104.

LUFTWAFFE "OFFIZIERSCHWERT" MODEL 1934

Blade: Finest quality hand-forged cast steel, polished and cleaned and heavily nickel-plated, then highly polished. The blade must be completely free of scratches and other marks or imperfections. When bent or impacted during testing, the blade must spring back exactly to its original form. The obverse side of the blade is etched with the maker's mark. Set against the crossguard on the ricasso, is a blue leather, oval washer. Weight of blade: approx. 310 grams.

Grip: Shaped wood, covered with blue Morocco leather, and bound with a twisted German-silver wire, heavily silver plated. Weight of grip including leather covering and ring: approx. 45 grams.

Crossguard: Cast German-silver (alloy: copper, nickel and zinc) with a nickel content of 15-16%, hand chiselled and finely polished; the relief-stamped center plate bearing a brass swastika is inlaid, and extra heavily silver plated. The crossguard is milled and the center plate inlaid and brazed on with a silver solder. Both wings of the crossguard are symmetrical.

Pommel: Cast German-silver (hollow-cast), nickel content 15-16% finely polished and heavily matt-finish, silver plated; swastika of brass. The center plates are relief-stamped and the pommel milled on both sides; the plates are inlaid and brazed on with a silver solder. The sides of the circular pommel are hand chiselled into an oakleaf pattern. Weight: approx. 95 grams.

Sheath: The sheath is made of a seamless aluminum tube, of the following specifications:

Alloying. . . . . .refined aluminum alloy
Tensile strength . . .28-32 kg per sq. mm
Stretch 10-15%

polished smooth and covered with blue Morocco leather, attached at the top. Thickness of leather is 1mm, maximum tolerance is 0.2mm or below. Weight of sheath including mounts and hanger: approx. 280 grams.

Sheath Mounts: (locket and drag) Sheet German-silver, with a minimum nickel content of 15-16%, heavily matt-silver plated.

Sheath Mount Screws: Of German-silver, with a minimum nickel content of 15-16%, matt-finish silver plated.

Blade Retaining Springs: Blued steel springs. The springs must be of sufficient strength to prevent the sword from moving side to side while in the sheath.

Hanger: Knob fittings and rings of German-silver, minimum nickel content of 15-16%, heavily matt-finish silver plated. Grained cow-hide leatherware, covered with blue Morocco leather, and hand stitched.

Measurements: Overall length - 92cm
Sheath length - 74.3cm
Sheath weight - 270 grams

Weight: Approximately 900 grams.

IN THE FOREGROUND, THE LUFTWAFFE "FLIEGERSCHWERT" IS BEING ASSEMBLED FOR PACKING AND SHIPMENT TO THE RETAIL OUTLETS.

LIBRARY OF CONGRESS

STATE FUNERAL FOR GENERALOBERST UDET WHO COMMITTED SUICIDE FOLLOWING LONG-TERM DISPUTES WITH MILCH, NOVEMBER 21, 1941. THE COFFIN WAS BORNE THROUGH THE STREETS WITH THE "FLIEGERSCHWERT" ATOP THE FLAG-DRAPED CASKET. ABOVE, REICHSMARSCHALL GÖRING PAYS TRIBUTE TO HIS DECEASED COMRADE. GENERALLEUTNANT ADOLF GALLAND, WITH HIS SWORD IN THE POSITION OF ATTENTION, IS AN HONORARY PALLBEARER. BELOW, FINAL TRIBUTE IS PAID TO UDET, HIS CASKET LOWERED INTO THE GRAVE STILL BEARING HIS SWORD.

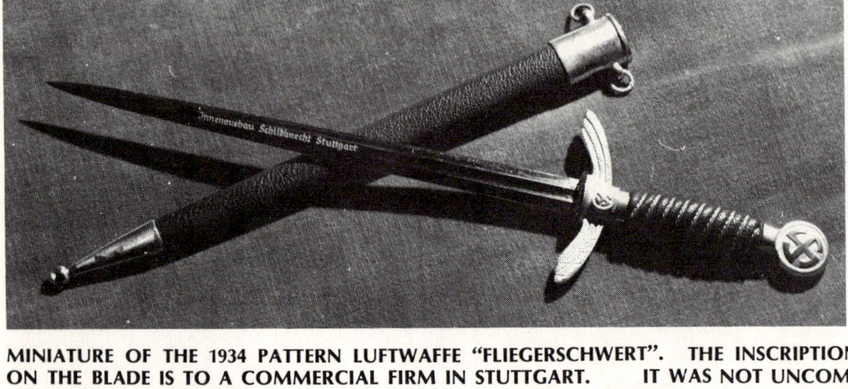

CAPT. R. J. COLE COLLECTION

MINIATURE OF THE 1934 PATTERN LUFTWAFFE "FLIEGERSCHWERT". THE INSCRIPTION ON THE BLADE IS TO A COMMERCIAL FIRM IN STUTTGART. IT WAS NOT UNCOMMON TO PASS THESE MINIATURES OUT FOR USE AS LETTER OPENERS AND DESK MOUNTS.

1st PATTERN LUFTWAFFE GENERAL'S "FLIEGERDEGEN". THIS THIN, STRAIGHT BLADED SWORD WAS DESIGNATED A "STICHDEGEN", AND WAS APPROVED FOR WEAR BY GÖRING. THIS PATTERN SWORD WAS IN USE FROM 1935 TO 1938.

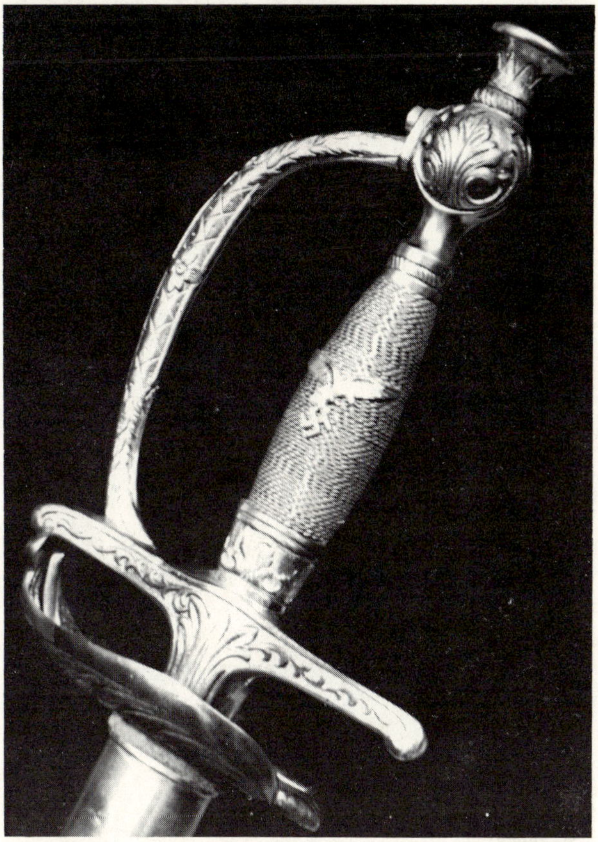

LIBRARY OF CONGRESS

THE LUFTWAFFE GENERAL'S "FLIEGERDEGEN" LIES ATOP THE CASKET OF GENERAL WEVER, LUFTWAFFE CHIEF-OF-STAFF, DURING HIS STATE FUNERAL ON JULY 6, 1936. A PHOTOGRAPHIC SERIES OF THIS, AND OTHER SUCH FUNERALS INDICATES THAT THE SWORD WAS BURIED WITH THE CASKET.

LUFTWAFFE GENERAL'S DEGEN PRODUCED BY THE EICKHORN FIRM. THE SPECIMEN SHOWN IS AN ORIGINAL PIECE. IT SHOULD BE NOTED, HOWEVER, THAT THIS IDENTICAL SWORD IS BEING PRODUCED BY EICKHORN TODAY, BUT THE GUARD BEARS THE CHILEAN AIR FORCE INSIGNIA, AND NOT THE LUFTWAFFE ONE. PROFIT-MINDED DEALERS HAVE REMOVED THE EXPORT INSIGNE AND REPLACED IT WITH A SILVER EAGLE OF THE LUFTWAFFE DESIGN. OBVIOUSLY, THE BLADE ETCHING IS ADDED TO THE PLAIN BLADE IN AN EFFORT TO CARRY THROUGH THE DECEPTION. THE WORK IS EXPERTLY DONE AND GREAT CARE SHOULD BE EXERCISED.

GARY WALKER MUSEUM COLLECTION

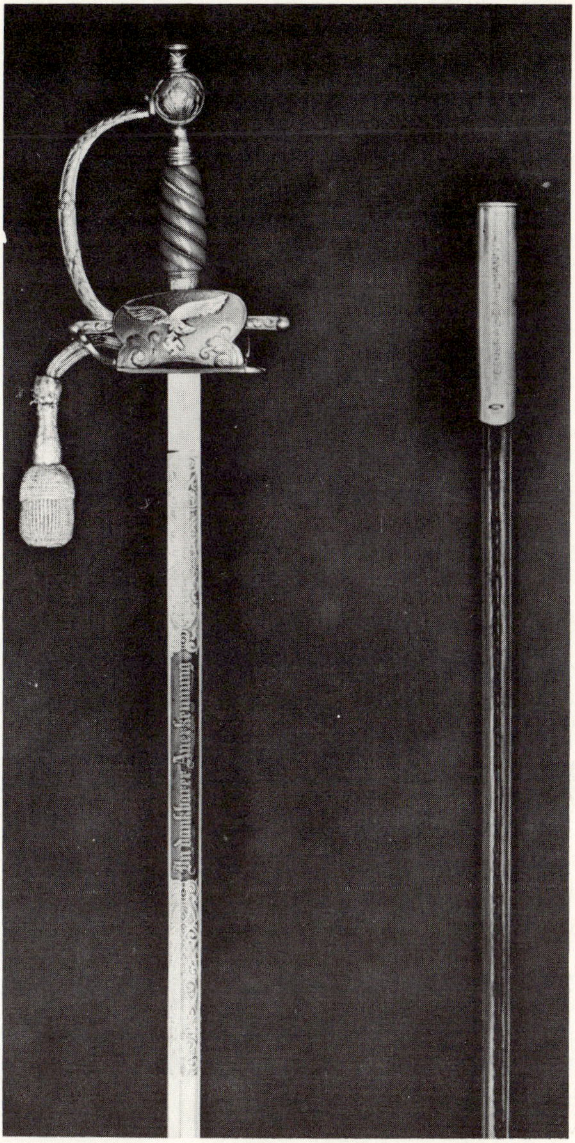

LUFTWAFFE GENERAL DAHL-MANN'S PRESENTATION "FLIEGERDEGEN" (NOTE FULL NAME ON THE REVERSE OF THE UPPER SCABBARD FITTING).

THE NAVY

The navies of the world are and have always been notoriously conservative, and the German Navy throughout its history was no exception. Largely professional in the execution of their duties, the officers and men of the German Navy were least likely to become politically involved even though they engaged in the Navy mutiny at the end of World War I, which resulted in the abdication of the Kaiser and the eventual peace which was achieved.

The Navy boasted the longest tradition of sword wear in modern Europe. The sword had long been an effective shipboard weapon, used in both offensive and defensive rolls, plus serving as an ornamental sidearm for officers on shore duty. The Navy sword, however, became a mere symbol of the previous history and a mark of rank after the turn of the century. The dress sword which was to evolve, had its inception in the "Kaiserliche Marine" prior to the advent of World War I, and underwent few changes from that time to the end of World War II. The transition from the Imperial sword to that carried by the officers of the Kriegsmarine saw only slight modifications. The Imperial crown was dropped from the design motif on the guard, and the etched blade pattern was also modified accordingly. The characteristic symbols of the Navy, the red and green eyes of the lionhead pommel and the fouled anchor were retained.

Prior to the Nazi era, there were no specific restrictions regarding the patterns of sword to be worn providing permission for wear was granted in those cases of the more varied styles. Precise regulations, however, did exist regarding the authorization for wear and the manner in which the sword could be worn. After Hitler formalized the military services in 1935, more stringent regulations were drawn up governing the wear of the "Säbel". Hitler immediately imposed his will on the military through a series of uniform regulations, revising the "Clothing and Lress Regulations for the Navy", in 1935 and again in 1937 (see translations of these regulations). Even though the Navy was forced to adopt the eagle and swastika along with the other military services, they continued to remain virtually aloof from the political intrigues which Hitler brought to the military.

As an outward display of the Navy's conservative mold, the Navy "Säbel" was very obviously void of any external display of the swastika, thus setting the Navy officer off from the other services as well as the governmental and political officials. That is not to say that there was no swastika incorporated into the etched blade design in various forms. Most common was the display of the swastika in the flags flown from the masthead of the ships or the use of the Wehrmacht "Adler" boldly setting off the etched design.

HITLER, ACCOMPANIED BY A LARGE DELEGATION OF HIGH-RANKING MILITARY AND POLITICAL OFFICERS, ATTENDS THE LAUNCHING OF ONE OF GERMANY'S POCKET BATTLE SHIPS. NOTE THAT THE TWO NAVAL OFFICERS (RIGHT FRONT) WEAR THE DAGGER AND SWORD WITH THE GREATCOAT AS PRESCRIBED BY NAVAL UNIFORM REGULATIONS.

Partial translation from "Clothing and Dress Regulations for the Navy", Naval Service Regulation 260, Berlin 1935... revised in 1937.

VI. The Naval Sidearm and how they are worn:

1. Sidearms for all ranks from Petty Officer upwards are: (a) the dagger, and (b) the sword.
2. Midshipmen carry the Naval dagger with portepee. Naval cadets carry the dagger without portepee.
3. Non-commissioned officers and ratings on duty wear the bayonet on a black leather belt when so ordered.
4. The dagger is suspended from two dagger straps and the Naval sword from two sword straps... both will be worn as follows: (dagger reference omitted)

(a) When the full dress belt is not worn:

With undress coat: The dagger or sword is worn suspended from the inside belt.

With frock coat: The dagger or sword is worn suspended from the inside belt.

With greatcoat: The sword is suspended from the inside belt.

With tropical dress: The sword is suspended from the inside belt.

With mess dress: The sword is not worn with the mess dress.

(b) When the full dress belt is worn:

With undress coat: The sword is worn suspended from the inside belt.

With frock coat: The sword is worn suspended from the inside belt.

With greatcoat: The sword is worn suspended from the inside belt.

With tropical dress: The sword is suspended from the inside belt.

When reporting, the Naval sword will be unhooked from the regain-hook and held beneath the basket with the left hand, fore and index fingers extended. The position of the sword must be perpendicular with the drag, approximately two inches above the ground.

If the greatcoat is worn, the lower left flap will be drawn back. When the sword is drawn, the left hand remains in position holding the scabbard.

On all other occasions, the sword may be worn in the regained (vertical) or unregained (diagonal) position. On occasions when military honors are to be rendered, the sword shall be worn in unregained position and held in the prescribed manner. If worn suspended from the outside belt, the sword is worn in the regained position but is still held by the scabbard in the above prescribed manner.

5. (Omit)

6. Those who served prior to November 11, 1918, in the War and were entitled to wear the sword or dagger with portepee, are permitted to wear the same Naval sidearm worn with honor during that period. These weapons may be worn unaltered, on or off duty, providing the present rank held corresponds to the weapon concerned, and that the occasion is appropriate for the wear of such pieces.

7-8. (Omit)

9. In foreign countries, the senior Naval officer afloat will establish the necessary restrictions regarding the wear of arms.

10. The sword portepee is strung around the upper part of the sword and then laced around the grip in a weave over the guard.

Dress Regulations for the Naval Field-Grey Uniform

XXVIII. Sidearms:

1. The Naval sword may be worn by personnel with the rank of Sergeant or Cadet and upward. With this type of uniform, the sword may be worn with an Army-type black, all metal scabbard instead of the usual leather scabbard.
2. The sword is worn suspended from the outside belt on two carrying straps when battledress or greatcoat is worn.
3. Naval Cadets and other ranks entitled to wear the portepee with their sword will also wear the portepee with their bayonet.

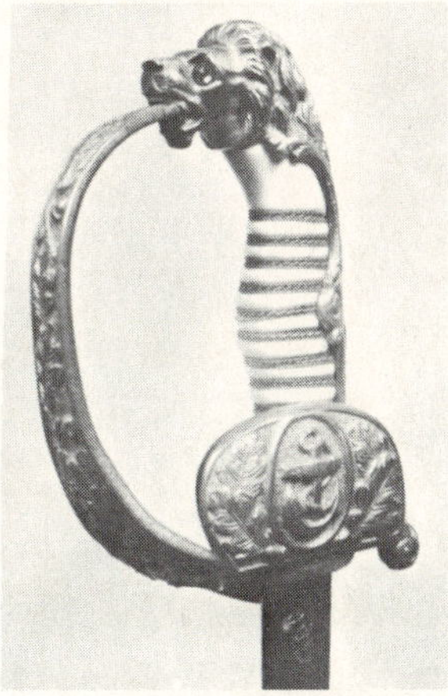

NAVY OFFICER'S SABER, WKC NO. 40. THE EYES OF THE LIONHEAD ARE RED AND GREEN INSET GLASS, SIGNIFYING THE "PORT" AND "STARBOARD" . . . CHARACTERISTIC OF THE SHIP'S RUNNING LIGHTS. THE PHOTO SHOWS THE FRONT GUARD SHELL IN THE UP POSITION. THE SHELL IS SPRING BALANCED AND FOLDS UPWARDS FOR PURPOSE OF STORAGE.

NAVY OFFICER'S SABER, EICKHORN NO. 1670, WITH IVORY GRIP. THE EICKHORN FIRM ALSO MANUFACTURED A NAVY OFFICER'S SABER IDENTICAL TO THE PATTERN SHOWN ABOVE AND WAS DESIGNATED PATTERN NO. 1513. NOTE THAT THIS PATTERN DOES NOT HAVE THE COLORED GLASS EYE INSETS.

NAVY ORDNANCE STAMP,
WEIMAR PERIOD

NAVY ORDNANCE STAMP,
1933-1938

REVERSE, SPRING-BALANCED LOCK GUARD AFFIXED TO UPPER SCABBARD FITTING OF THE NAVY OFFICER'S SABER. NOTE THE NAVY ORDNANCE STAMP AND CONTROL NUMBERS. THE CONFIGURATION OF THIS STAMP WOULD PUT THE DATE OF THE PIECE FROM 1938-1944.

THE "LAH" MARCHING PAST HITLER ON HIS 50th BIRTHDAY, APRIL 20, 1939.

Swords: SS-Police-Fire Dept.

THE SS

These organizations have been grouped together since they were in effect of the same subordination. In the decree of June 17, 1936, Himmler was accorded full control over the Police (to include the Fire Protection Police) upon his appointment as Commander-in-Chief of the German Police system. Since virtually all of the Police and civil defense organizations for all of Germany were placed under Himmler's control, administration and control responsibilities fell to the SS. As a result, after 1936, the basic sword pattern of the SS, Police and Fire Department were identical. The only difference was in the grip inlay which distinguished the SS from the Police/Fire Department.

From 1934 to 1936, officers of the SS carried a "Säbel" similar in design to the Army officers' saber, except that the hilt was heavily plated in old silver. During this period, there was some divergence in pattern design which would probably account for the adoption of a single pattern in 1936. The "SS-Ehrendegen" (Honor Sword) was awarded to all persons designated honor graduates of the SS Officers' Training School, or who were singled out by the Reichsführer-SS for exceptional service. The addition of the SS motto, "Meine Ehre heisst Treue", etched on the obverse of the blade was the only feature which distinguished the "Ehrendegen" from the standard pattern "Führerdegen".

Mounted enlisted personnel were issued the standard pattern Army cavalry saber, which has a plain nickel hilt, plain curved blade, and black painted metal scabbard.[1] In a few cases, the obverse of the langet was deeply engraved with the "SS-Sigrunen".

The "SS-Führerdegen" was an item of award, presented by the Reichsführer-SS to senior ranking officers of the SS and presented at the time of graduation to company grade officers upon completion of the SS Officers' School. If lost, it was necessary for the bearer to present

---

[1] Das Schwarze Korps, dated 10. 6. 37, Vol. #2.

his award certificate, and purchase a replacement from personal funds. Unlike military officers, the SS officer was required to order his replacement sword through the SS Clothing Office channels.

When a new pattern sword, the "SS-Führerdegen", was adopted in 1936, the practice of bestowing the Honor Sword continued. Again, the sole distinguishing feature was the addition of the SS motto to the blade. By 1937, the old pattern "SS-Führersäbel" had been phased out. By the end of 1937, only 1,442 "Ehrendegen" had been awarded.[2]

All officers and SS NCOs (SS-Portepeeführer) were authorized to wear the sword knot affixed to the hilt.

1934-1936 "SS-FÜHRERSABEL" PRODUCED BY THE EICKHORN FIRM. IN THIS PARTICULAR CASE, THE SPECIMEN IS THE HONOR SABER (EHRENSÄBEL). THE HILT FITTINGS ARE HEAVILY SILVER-PLATED. THE EYES OF THE LIONHEAD POMMEL ARE GREEN GLASS. THE BLADE BEARS THE SS MOTTO, WITH NO ETCHING ON THE REVERSE OF THE BLADE. THE SCABBARD IS THE STANDARD BLACK PAINTED STEEL SABER PATTERN.

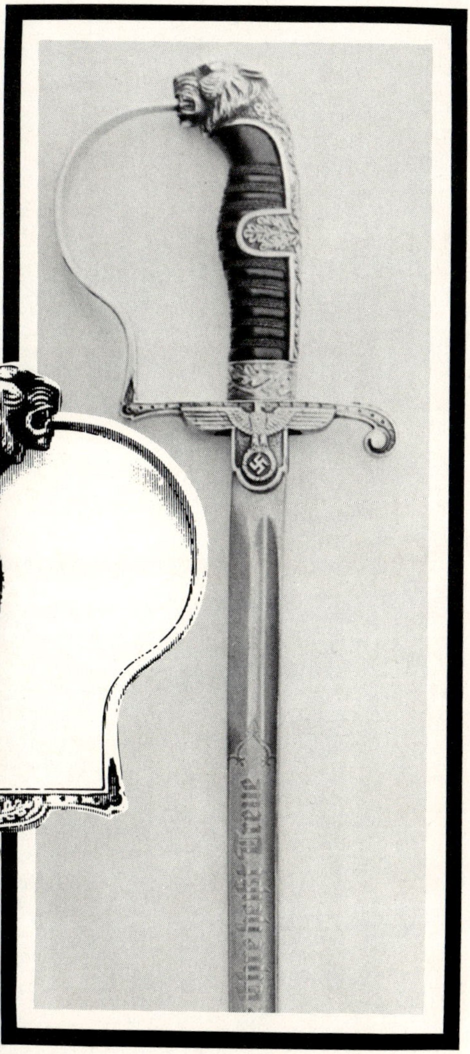

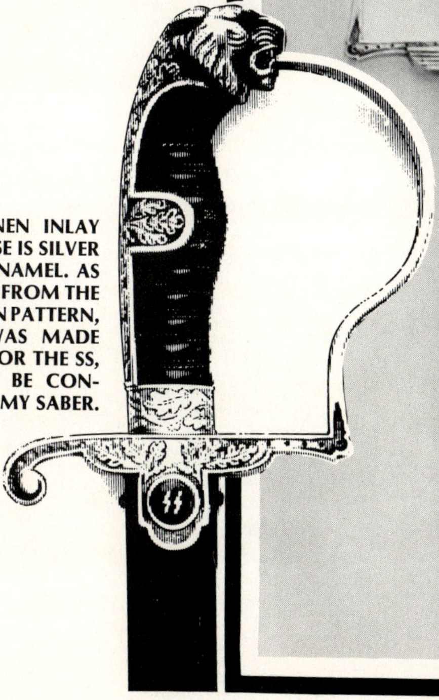

THE SS-SIGRUNEN INLAY ON THE REVERSE IS SILVER WITH BLACK ENAMEL. AS CAN BE NOTED FROM THE REVERSE DESIGN PATTERN, THIS SABER WAS MADE SPECIFICALLY FOR THE SS, AND CANNOT BE CONSIDERED AN ARMY SABER.

[2] SS-Dienstaltersliste der Schutzstaffel der NSDAP, Stand vom 1.12.37.

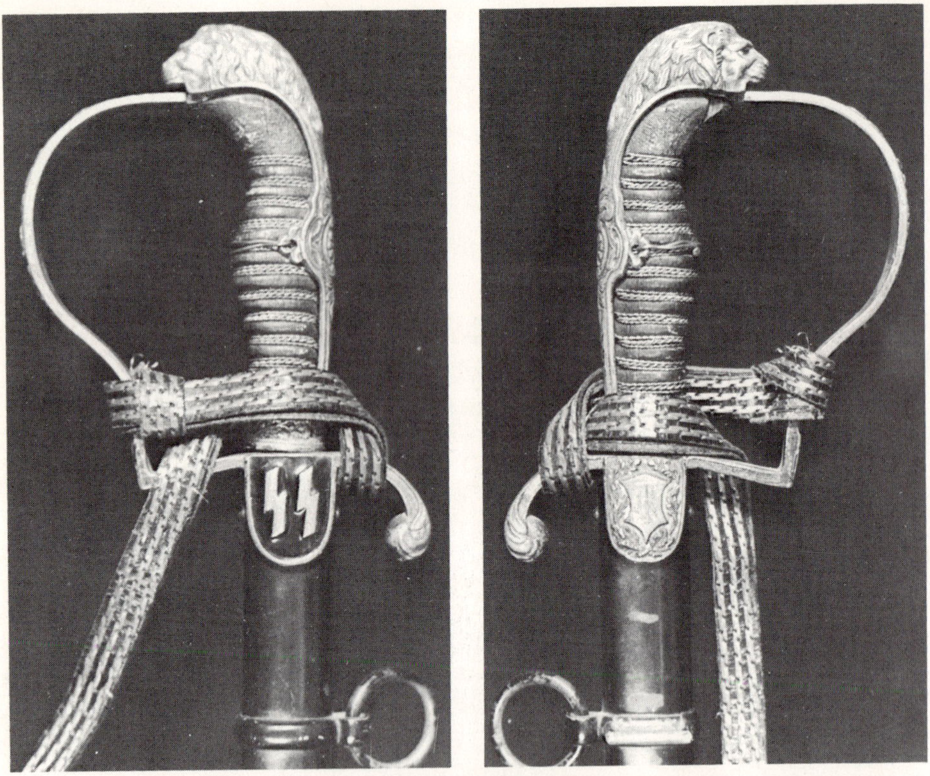

GEORGE R. CANADAY COLLECTION

A VERY EARLY SS OFFICER'S SABER INCORPORATING THE SS-SIGRUNEN IN THE LANGET. THIS IS JUST ONE OF THE MANY UNOFFICIAL PATTERNS WHICH EXISTED UNTIL 1936, AND ALTHOUGH NOT MARKED, THE PIECE WAS PROBABLY MADE BY THE P.D. LÜNESCHLOSS FIRM.

AT LEAST THREE DIFFERENT PATTERNS OF THE EARLY SS SABER ARE DISTINGUISHABLE IN THE RANK OF SS OFFICERS ABOVE.

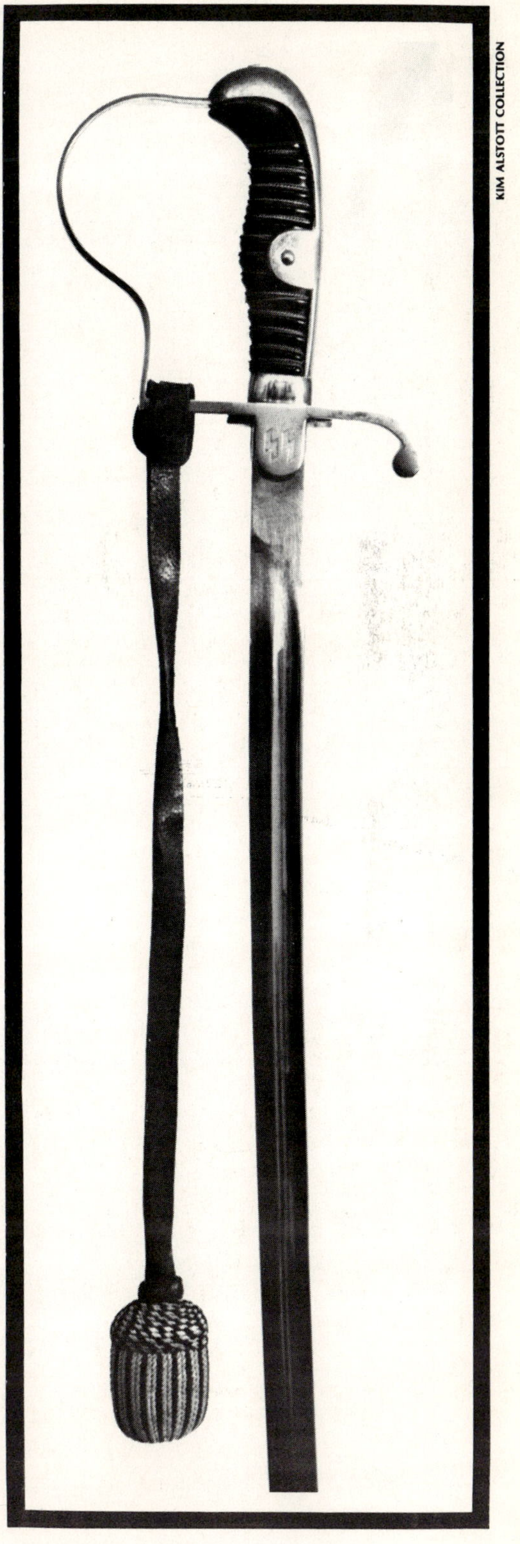

ARMY CAVALRY PATTERN SABER CARRIED BY MOUNTED PERSONNEL OF THE SS. THIS SPECIMEN HAS THE SS-SIGRUNEN DEEPLY CAST IN THE RICASSO. THE HILT IS NICKEL-PLATED WITH BLACK CELLULOID-COVERED HANDLE. AFFIXED IS THE REGULATION KNOT FOR SS-PORTEPEEFÜHRER.

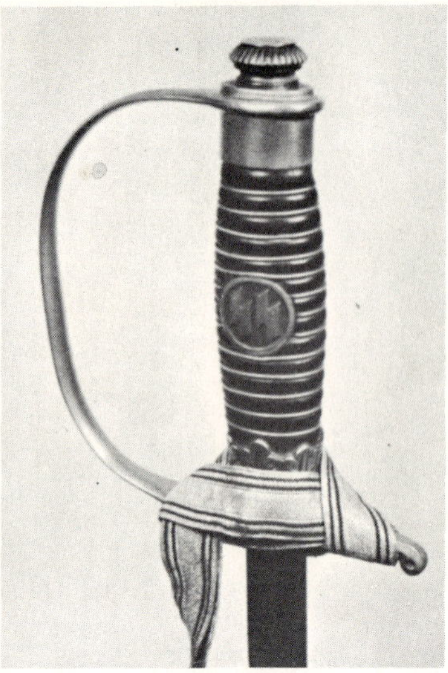

SS-FÜHRERDEGEN (SS OFFICER'S SWORD). THE HANDLE IS BLACK PAINTED WOOD WITH WIRE GRIP. THE HILT AND BOTH SCABBARD MOUNTS ARE NICKEL-PLATED. THE SCABBARD IS STRAIGHT TUBE STEEL WITH AN UPPER MOUNT AND LOWER SHOE.

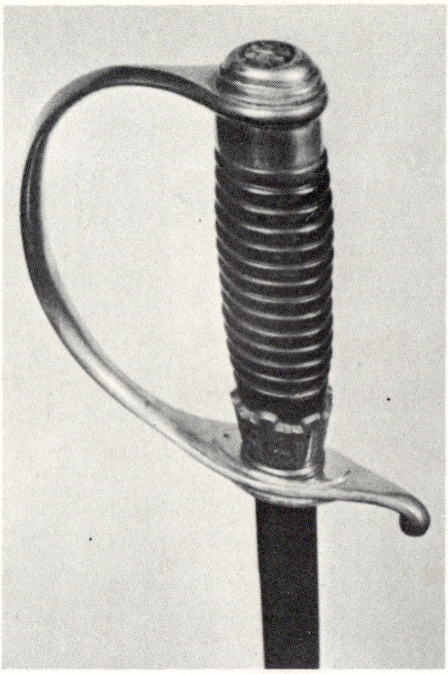

SS-PORTEPEEFÜHRERDEGEN (SENIOR SS NCO SWORD). NOTE THE SS-SIGRUNEN IN THE POMMEL LOCK NUT. THE HANDLE IS BLACK PAINTED WOOD. METAL HILT MOUNTINGS AND UPPER SCABBARD MOUNT ARE NICKEL-PLATED. THE SCABBARD DOES NOT HAVE A LOWER SHOE, BUT DOES INCORPORATE THE SCABBARD DRAG.

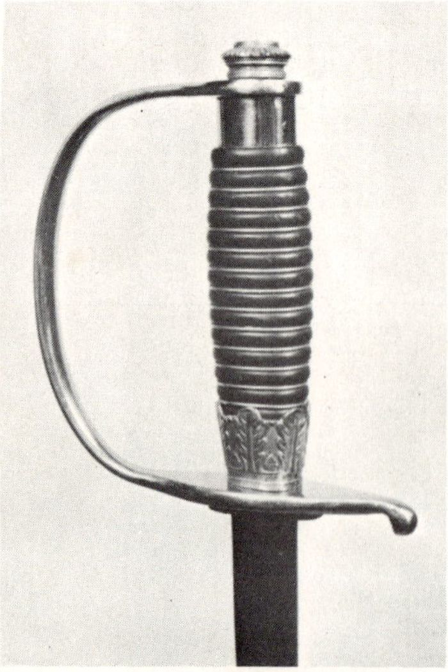

SS-BEWERBERDEGEN (OFFICER CANDIDATE'S SWORD), LEADER PATTERN. SAME AS THE OFFICER'S SWORD, BUT WITHOUT THE SS INLAY.

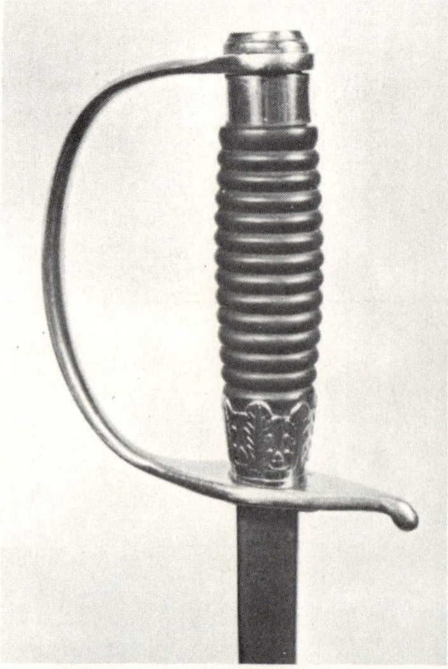

SS-BEWERBERDGEN (OFFICER CANDIDATE'S SWORD), SUBORDINATE PATTERN. NO SS-SIGRUNEN CAST IN THE POMMEL LOCK NUT.

SS PERSONNEL SWEARING THE OATH OF ALLEGIANCE TO ADOLF HITLER.

"What is the tenor of thy oath?"

"We swear to thee, Adolf Hitler, Führer and Chancellor of the German Reich, to be loyal and brave. We swear to thee, and to the leaders whom thou mayst designate, obedience unto death. So help us God."

The oath was sworn on an object dear to the unit, in this case the blade of the SS sword. Another example would be to touch the hull of a tank or the bunting of the Company standard, etc.

THE REICHSFÜHRER-SS BESTOWING THE "EHRENDEGEN" ON SS OFFICER CANDIDATES.

HOOVER LIBRARY, STANFORD UNIVERSITY

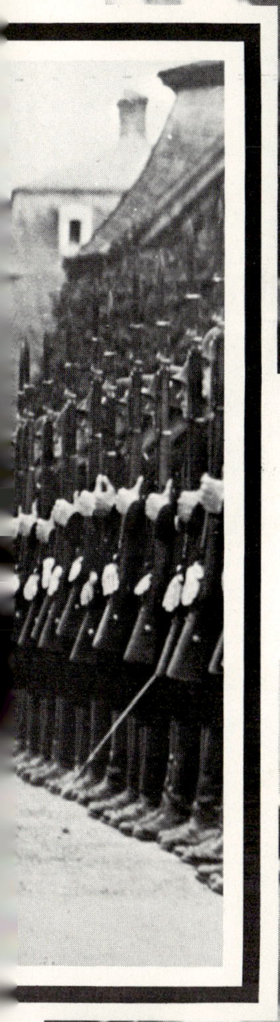

REICHSFÜHRER-SS HIMMLER AT GODESBURG, MARCH 28, 1936.

POLICE/FIRE DEPARTMENT

Like the SS, both the Police and Fire Department officials were required to wear a saber of Army design until their distinct pattern sword was introduced in 1936. There was no fixed pattern up to this time, thus a wide variety of their early saber can be encountered. After the introduction of the 1936 pattern, however, all swords and their wear were standardized. There was a single departure in the pattern swords worn by the Fire Protection Police. Persons holding the rank of Wehrführer and Hauptbrandmeister upwards to officer rank, were entitled to wear the Army pattern sabers, "Blücher", "Wrangel" and a plain hilted "Säbel" with a gold colored, metal hilt with black celluloid handle. All officers with the rank of Oberbrandmeister upwards had the option of wearing the "Polizei-Führerdegen" or the Fire Department dress sidearm. Contrary to previous publications, subordinates of the Fire Department were not authorized to wear the "Polizei-Unterführerdegen".

By 1944, wear of the sword had given way to the pistol sidearm when performing outside duty. Wear was continued (especially by the SS), however, at formal functions and official ceremonies.

(Partial translation from a 1943 Police Manual)

BLADE SIDEARMS OF THE POLICE

There are the following types of edged weapons:

(a) Short edged weapons (bayonets), some of which can be affixed to rifles.
(b) Long edged weapons (sabers and swords). Sabers have a slightly curved blade, while sword blades are straight.

The Police are equipped with the following edged weapons:

(a) S84/98; S98/05; Police bayonet, affixable (P. S. a.);
(b) Police bayonet, not affixable (P. S. n.);
(c) Police bayonet, special type (P. S. S.);
(d) Saber for mounted personnel (S. f. B.);
(e) SS-Police sword (officer and NCO swords).

Police Sword (Polizeidegen):

Officers

Metal portions of the hilt are nickeled steel. Handle is black-dyed wood with Leichtmetall (aluminum) wrapping. The Police insigne is convex and made of silvered aluminum. Ferrule is chiselled, nickeled and lightly silvered (old silver finish...oxidized). Scabbard mouth piece is same finish as ferrule as is the bottom mount with small drag.

NCOs

Differs in that:

(a) Knuckle bow (guard) is somewhat heavier.
(b) Pommel button is flat and oval.
(c) Grip is without wire grip. Insigne is same as that on the Police bayonet, but old silver color.
(d) Bottom mount not present on sheath. Instead, the drag is mounted directly on the sheath as with the saber.

EQUIPPING WITH SIDEARMS

1. Worn are:

   (a) SS-Officer's Sword by offiziere, Meister and Oberjunker of the Orpo, who are given it by the RF-SS.
   (b) SS-Police Sword (Officer's)...those ranks listed in (a), but not entitled to carry the SS-Officer's Sword.
   (c) SS-Police Sword (NCOs) by Hauptwachtmeister rank. Additionally, it may be worn by:
      (1) Revier-(Bezirks)-Oberwachtmeister and Zugmeister ranks.
      (2) Officials of the Orpo from Oberwachtmeister upward after a total service time (both military and police) of nine years.
      (3) Officials of the Orpo from Oberwachtmeister upwards, who as former members of the Wehrmacht or Landespolizei were entitled to wear the officer's sidearm.
   (d) The S.f.B. by mounted Wachtmeister (SB.) not entitled to carry the sword under (a) thru (c).
   (e) Police bayonet or S84/98 or S98/05 by all Wachtmeister not entitled to wear the sword or saber under (a) thur (d).

2. Officer's swords in 1(a) thru 1(b) can be worn with any uniform, or by Orpo Meisters, but only with abbreviated Service thru Dress and Walking-Out uniforms; with uniform for action by Police, the S84/98 is carried in place of the sword by dismounted personnel. The Police bayonet, S84/98 or S98/05 is worn with the Service Uniform(1) for unmounted Meister of Gendarmerie.

3. The NCO sword in 1(c) is worn by the Senior Company or Squadron NCO with all types of uniforms. The S84/98 is worn in place of the sword for unmounted police action.

[1] Not worn during action by Orpo and not with Service uniform of the Gendarmeriemeister on the beat. Further, swords may not be worn by Gendarmeriemeistern at closed activities of the Gendarmerie (e.g. funerals), since the Meister is then designated as a Zugführer.

4. Senior NCOs (not the leading Senior NCO), also officials from Revier-(Bezirks)-Oberwachtmeister up named in 1(c), (1)-(3) may only carry the NCO sword with abbreviated Service thru Dress and Walking-Out Dress uniforms.

5. Officers and officials of the motor service (including motorized Gendarmerie and motorized traffic units) carry no sidearm except the pistol when on motor duty.

6. The S. f. B. in 1(d) is carried only on horseback duty or training duty. On any dismounted duty, the sidearm for mounted officials is provided. Mounted officials will carry the S. f. B. when wearing the Dress, Walking-Out Dress and Off-Duty uniforms.

DELIVERY OF SIDEARMS

1. From state resources without payment for use (item of issue):
   (a) SS-Police Sword (officer's), but only Meisters to whom it is not presented. (2)
   (b) SS-Police Sword (NCO), but only for leading Senior Company or Squadron NCO. (3)
   (c) Sabers for all mounted officials not included under (a) or (b).
   (d) Bayonets for all officers and police officials required to carry same for duty.

2. Until officially replaced with SS-Police Swords (officer's and NCOs), those officials in 1(a) and 1(b) may wear the earlier Police Officer's Saber.

3. SS-Police Swords (officer's and NCOs) and SS-Officer's Swords not provided by 1(a) and 1(b) from state resources, will be privately purchased by those authorized them.

WEAR OF SIDEARMS

1. Swords and the old Police Saber are worn on the hanger outside the coat or topcoat:
   (a) With Parade uniform, Mess Dress and pertinent uniform types to which the dress hanger with trimming of spun aluminum (spun golden-yellow metal for general officers) goes thru the slit in the pocket of the coat or topcoat.
   (b) With all other uniform types on the dress hanger without trimming worn thru the pocket slits of the coat and topcoat.

2. The leather hanger for 1(a) is the color of the tunic; leather for 1(b) is the color of the leather items, usually black (brown for Gendarmerie).

   The S. f. B. is worn by mounted officials on the outer belt with saber cross-strap. With Dress and Walking-Out Dress and Off-Duty, the saber may be worn thru the pocket slits of the coat and topcoat on the leather (Gendarmerie brown, others black) hanger suspended from the inner belt.

   On horseback, the swords (former Police Officer's Saber and S. f. B.) are mounted on the saddle.

   Bayonets are carried in the bayonet frog on the outer belt with all uniforms.

   Swords and sabers are to be left on the carrying loop (regained position) at parades, when reporting, etc., and the saber sheath held upright with the left hand as described.

PHOTO: MARC BROWN/HICKOX COLLECTION

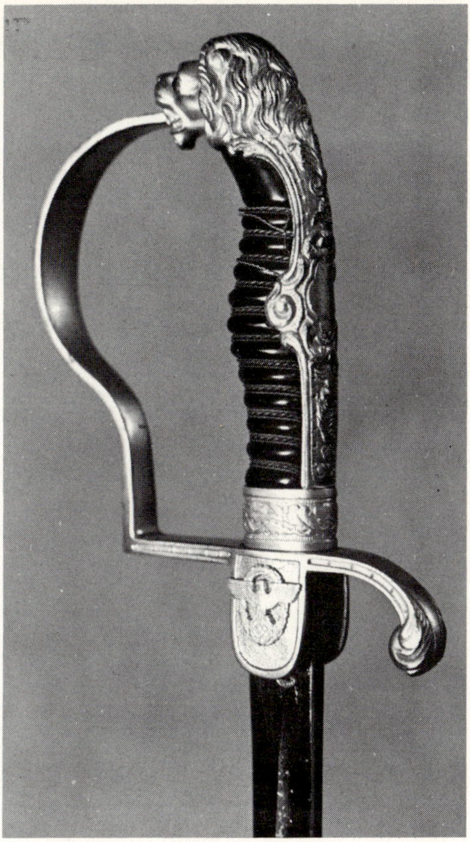

VERY EARLY WKC POLICE OFFICER'S SABER. THE HILT MOUNTINGS ARE BRASS WITH THE ALUMINUM NATIONAL POLICE INSIGNE AFFIXED TO THE LANGET.

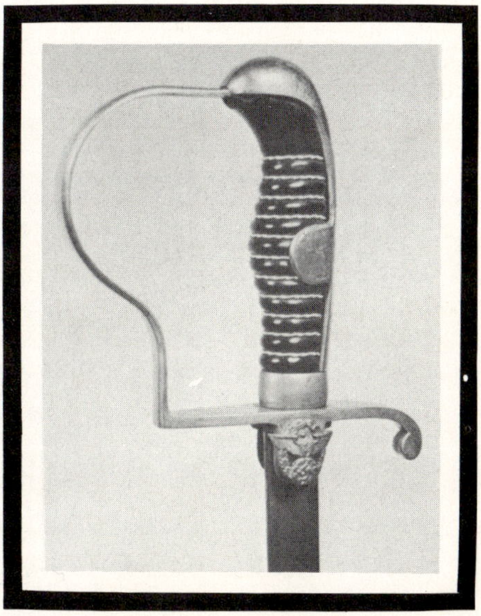

STANDARD PATTERN OFFICER'S CAVALRY SABER, BUT WITH THE BRONZE POLICE INSIGNE AFFIXED TO THE LANGET. THIS MODIFIED PATTERN AND THE OTHER EARLIER UNOFFICIAL PATTERNS GAVE WAY TO THE OFFICIALLY DESIGNATED "POLIZEIFÜHRERDEGEN", IN 1936.

---

[2] Officer's Swords were not being made during the war and also not being bestowed.

[3] NCO Swords were not being made during the war.

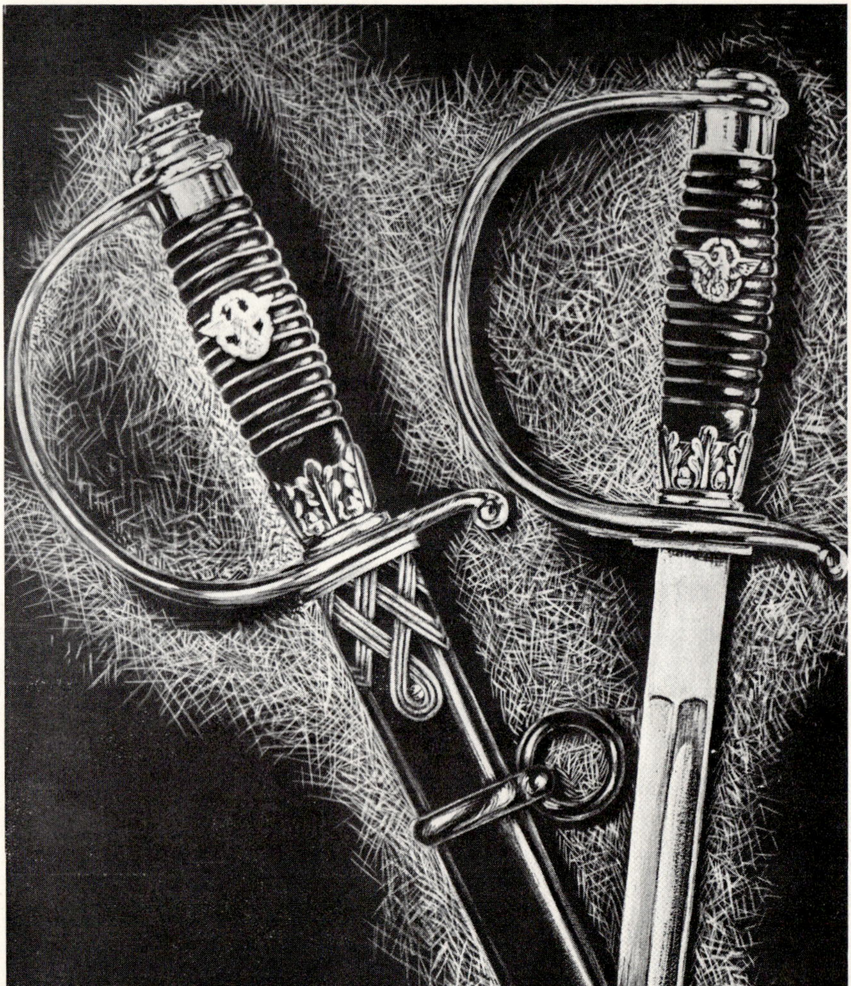

(LEFT) POLICE "FÜHRERDEGEN". IDENTICAL TO THE "SS-FÜHRERDEGEN" WITH THE EXCEPTION OF THE BRONZE NATIONAL POLICE INSIGNE INLAY ON THE FACE OF THE HANDLE. AFTER 1937, THE UNDERSIDE OF THE KNUCKLE BOW AND/OR THE UPPER SCABBARD THROAT MIGHT BE STAMPED WITH THE OVERLAPPING SS-SIGRUNEN TO SHOW THAT THEY HAD BEEN INSPECTED BY SS INSPECTORS AT THE TIME OF MANUFACTURE.

(RIGHT) POLICE "UNTERFÜHRERDEGEN" FOR WEAR BY SENIOR NCO'S. THE NATIONAL POLICE INSIGNE INLAY IS ALUMINUM. ALL METAL HILT AND UPPER SCABBARD FITTINGS ARE NICKEL-PLATED.

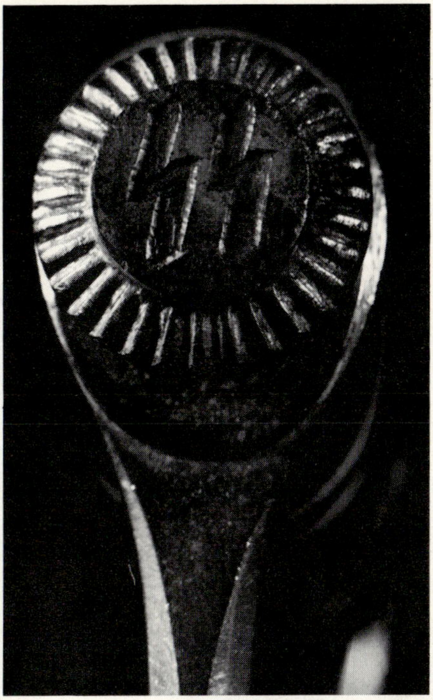

SS-SIGRUNEN DEEPLY ENGRAVED IN THE POMMEL LOCK NUT OF THE "SS-POLIZEIFÜHRERDEGEN. THIS WAS NOT A NORMAL PRACTICE AND WAS PROBABLY DONE AT THE REQUEST OF THE WEARER AT OPTIONAL COST. THE SAME CAN BE SAID OF THE OWNER'S INITIALS, WHICH WERE ALSO OFTEN PLACED IN THIS SPACE.

R. MUNDHENK COLLECTION

THE POLICE ON PARADE IN BERLIN.

AS LATE AS 1944, HAUPTMANN ALOIS TRESS WORE NOT THE REGULATION POLICE "FÜHRERDEGEN", BUT A STANDARD ARMY OFFICER'S SABER. INSTEAD OF HAVING DUAL MEMBERSHIP IN THE POLICE AND SS AS MANY OF HIS POLICE COMRADES, HAUPTMANN TRESS WAS A LIEUTENANT IN THE FELDGENDARMERIE OF THE ARMY DURING THE WAR YEARS. STATING THAT HE OWNED BOTH THE ARMY OFFICER'S SABER AND THE POLICE "FÜHRERDEGEN", HE HAD THE OPTION OF WEARING EITHER.

UNATTRIBUTED EARLY FIRE DEPARTMENT OFFICIAL'S SABER. NOTE THE USE OF SHARK SKIN ON THE HANDLE, WHICH WAS DISCONTINUED AFTER 1930. THIS SPECIMEN HAS AN ETCHED BLADE.

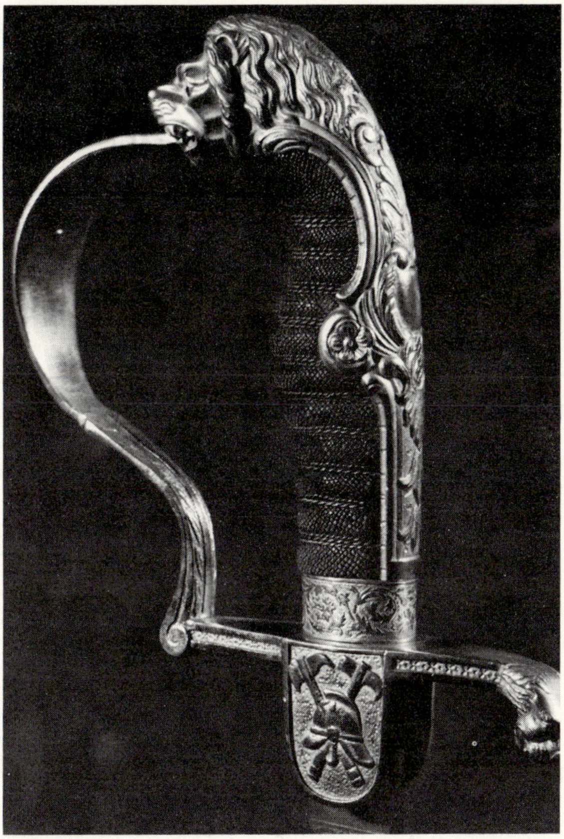

CAPT. R. J. COLE COLLECTION

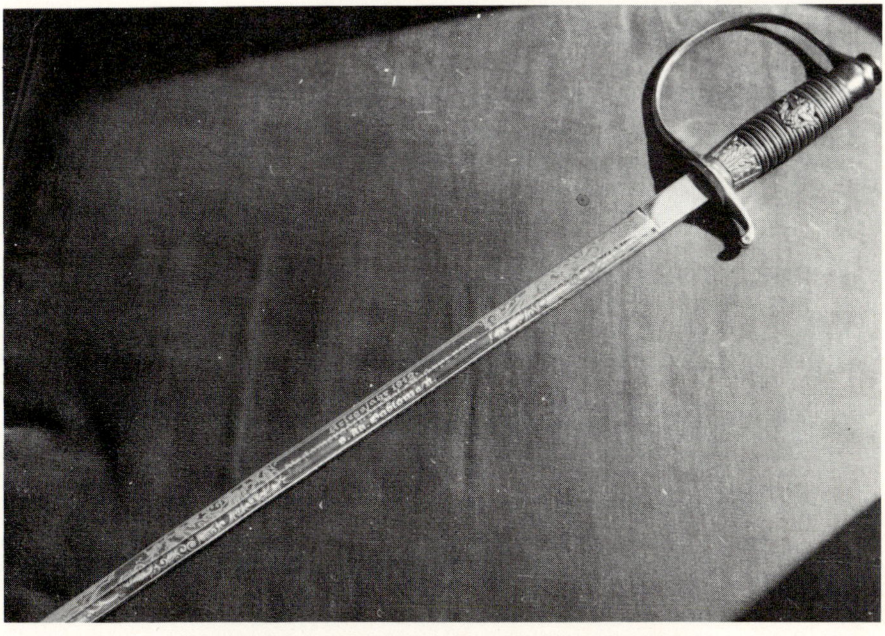

FIRE DEPARTMENT OFFICIAL'S PRESENTATION SWORD.

Swords: Miscellaneous

HITLER AND HIS DIPLOMATIC DELEGATION IN ROME, MAY 1938.

CUSTOMS AND RAILWAY PROTECTION SERVICES

The Reich Finance Administration Service had direct control over the Land Customs (Landzoll), Water Customs (Wasserzoll) and the Railway Protection Service (Bahnschutz). Services performed by the three elements resulted in a great deal of annual income for the national treasury as well as providing security for the national borders. Since most of the functions performed were of a police type, these elements worked closely with the police and SS.

Sword-bearing officials had a limited selection of sword patterns from which to choose, and more often than not, opted to wear the Army pattern "Säbel". The "Blucher" and "Freiherr von Stein" were two such Army patterns authorized for wear. At least two firms... Eickhorn and WKC, however, designed a saber specifically for the "Zoll".

Prior to the outbreak of the war, officials of the Water Customs wore the Navy Officer's "Säbel" with portepee. A distinctive "Säbel", however, was created for this element as well.... a lionhead pattern with a gilt finish. The scabbard was the main distinguishing feature which set it apart from a standard Army pattern, being virtually identical to the Navy scabbard. The scabbard was mounted with an upper and center fitting with swivel rings for suspension, and the leather was colored blue/black. A brass shoe with drag protected the tip of the scabbard.

No distinctive pattern saber was ever designed for the officials of the Railway Protection Service. They were authorized to wear the Army pattern saber, however, specifically the "Wrangel" and "Roon".

Subordinate personnel of the three elements not authorized to wear the "Säbel" were required to be armed with the standard service bayonet. Members of the Land Customs wearing the bayonet also wore a distinctive bayonet knot affixed to the suspension frog. The suspension strap was identical to that worn on the "Säbel" (grey-green) and the acorn portion was a pattern of two silver and one green twist braid in alternating rows. The bayonet knot for the Water Customs was identical except that the acorn was two gold and one blue twist cord in alternating patterns.

PHOTO: ANDY WALKER

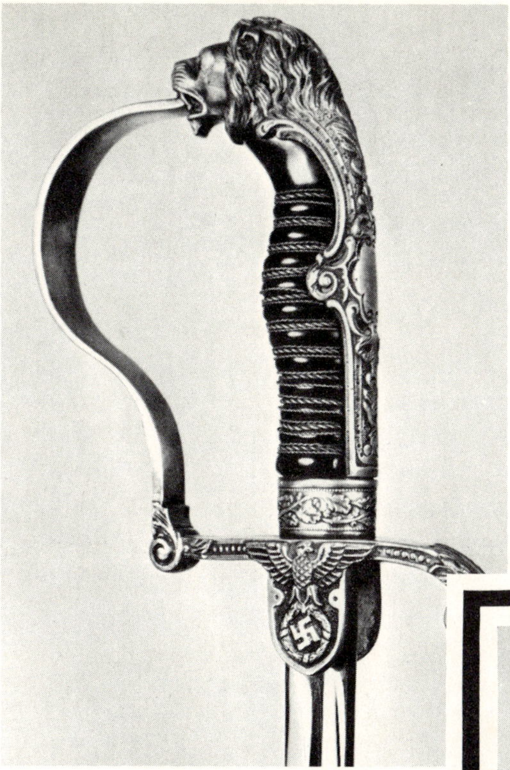

WKC LAND CUSTOMS SABER NO. 1016, SILVER HILT WITH GREEN EYES.

The Imperial-period, Bahnschutz Gala-Degen at right was manufactured by the Eickhorn Firm. The hilt was constructed of solid brass, and disassembled into five parts. The blade was finely etched. The leather scabbard (black) was mounted with ornately designed upper and lower brass fittings. The purpose of showing this sword is to emphasize that such a pattern did not exist for wear during the Nazi era, but sometimes can be found with a swastika motif added to the design.... usually at some point in time after the war's end.

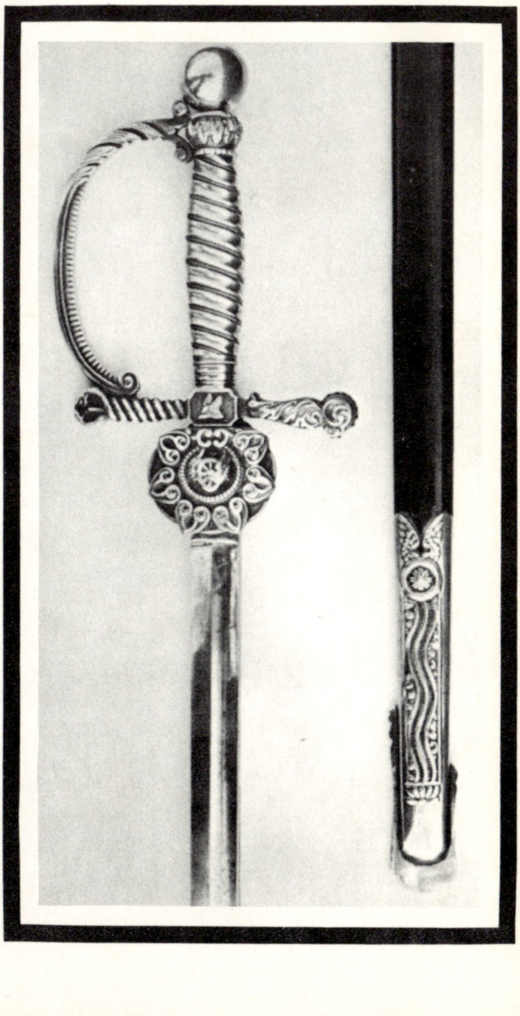

DIPLOMATIC SERVICE

The hilt mountings were heavily plated in old silver, and the grip side-plates were made of celluloid simulating mother-of-pearl on the Alcoso Diplomatic Official's Sword, No. 1501, below. The national eagle was identical in design to that which decorated the Reichstag. Approved in 1938, the sword was worn with the formal dress uniform, and adorned with the Diplomatic portepee. It can be assumed that the swords of the highest ranking officials (and very definitely, Ribbentrop) were finished with a gold plate. The straight double-edged blade was ornately

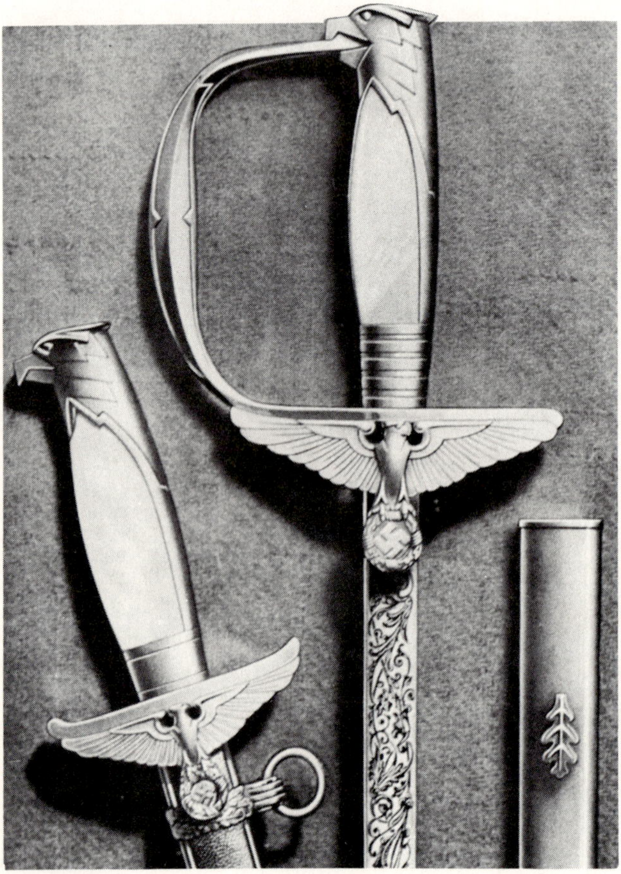

etched with a floral design, and the national eagle served as the central design feature. The scabbard was leather with metal scabbard throat and shoe.

Also shown is the Diplomatic Official's Dagger, No. 1502. Note that the eagle's head on the crossguard of both the sword and the dagger face in the opposite direction from the eagle head pommel. This is the only feature which sets the Diplomatic Official's Dagger off from the Government Official's Dagger.

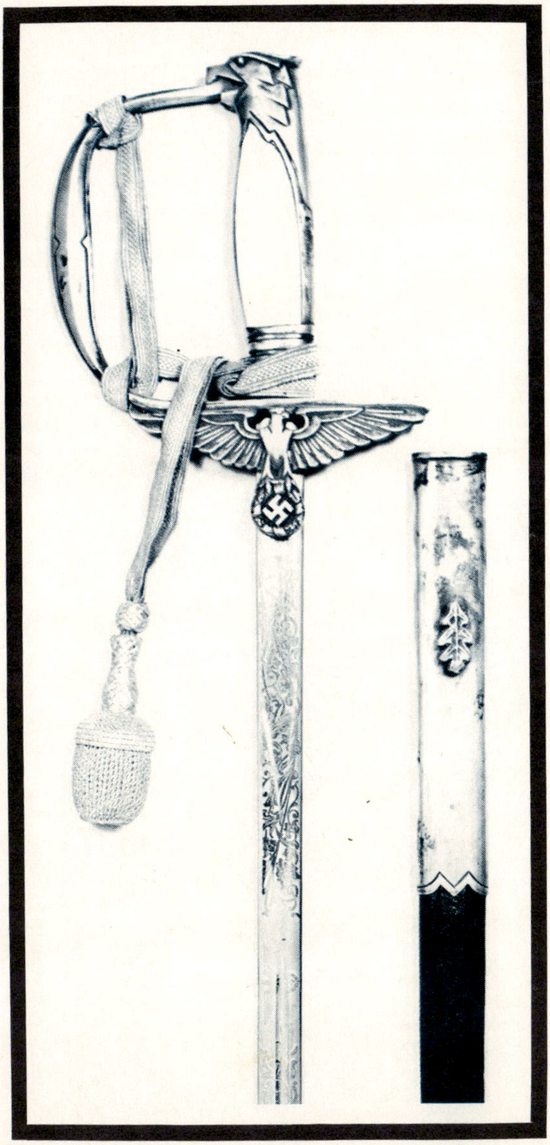

OBVERSE OF THE DIPLOMATIC OFFICIAL'S "DEGEN" WITH PORTEPEE. NOTE THE NATIONAL EMBLEM ON THE OBVERSE BLADE PATTERN.

JUSTICE AND PRISON ADMINISTRATION

Higher officials of the Justice Administration Service (Justizbeamte) were authorized to carry the pattern "Säbel" below with a gold wash finish. Members authorized to carry a saber in the Prison Administration Service (Strafsvollzugsbeamte) carried an identical pattern piece, but finished in white aluminum or old silver. This is the only pattern known whose design on the obverse langet is identical with the design of the reverse.

A very few reproductions of this pattern saber have been expertly crafted from solid brass, and are new in their entirety.

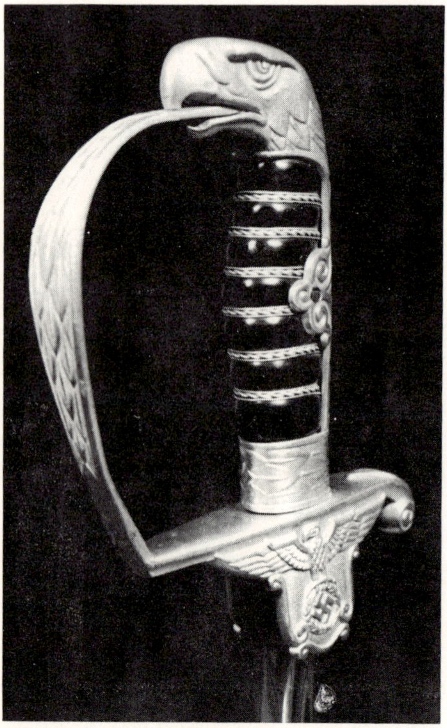

EICKHORN JUSTICE OFFICIAL'S SABER NO. 1763.

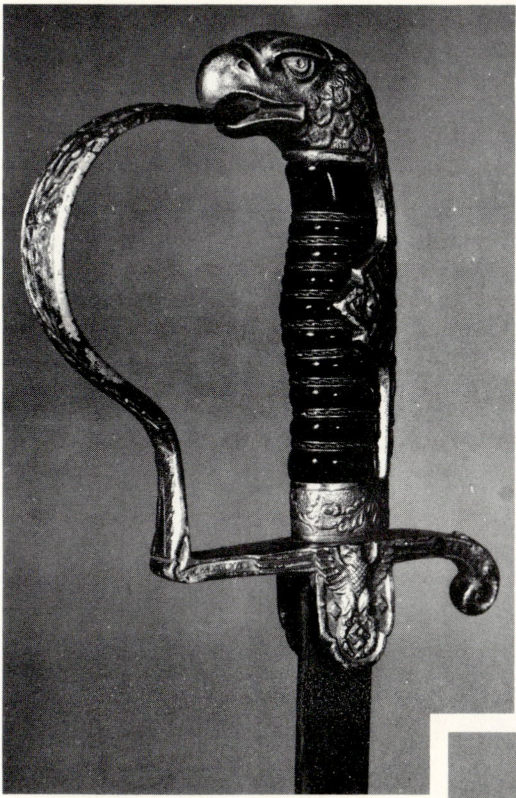

PHOTO: MARC BROWN/HICKOX COLLECTION

F.W. HÖLLER UNATTRIBUTED PATTERN, PROBABLE JUSTICE OFFICIAL'S SABER.

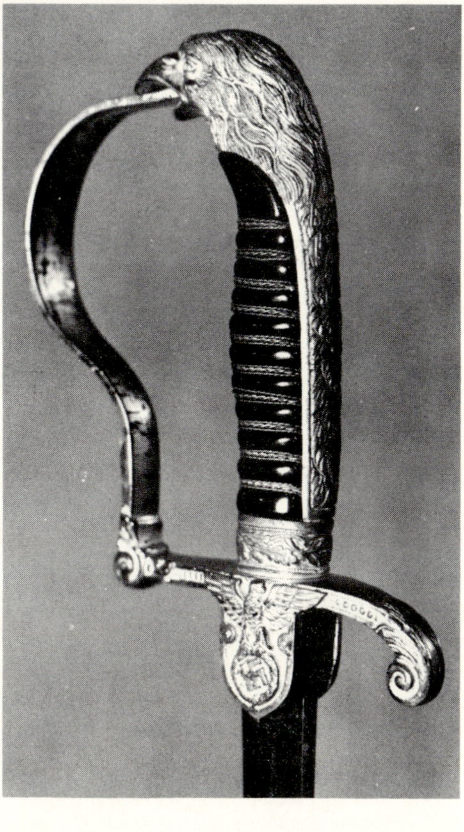

PHOTO: MARC BROWN/HICKOX COLLECTION

WKC JUSTICE OFFICIAL'S SABER NO. 1047, SILVER HILT WITH GREEN EYES.

BERGBAU (MINER'S) SERVICE

One of the oldest pattern swords, dating back prior to the turn of the century, is the Bergbau (Miner's) "Säbel". Pre-Nazi era sabers are normally found with the sharkskin covered handle whereas those of the Nazi era are black celluloid over wood, with a wire wrap. Various patterns of the Miner's saber are manufactured, and no single piece can be designated as the official pattern. All attributed to the Bergbau, however, have the crossed hammers incorporated somewhere in the design motif.

GARY WALKER MUSEUM COLLECTION

BERGBAU(MINER'S) SWORDS WITH DOUBLE ETCHED BLADES.

MISCELLANEOUS

GARY BARNUM COLLECTION

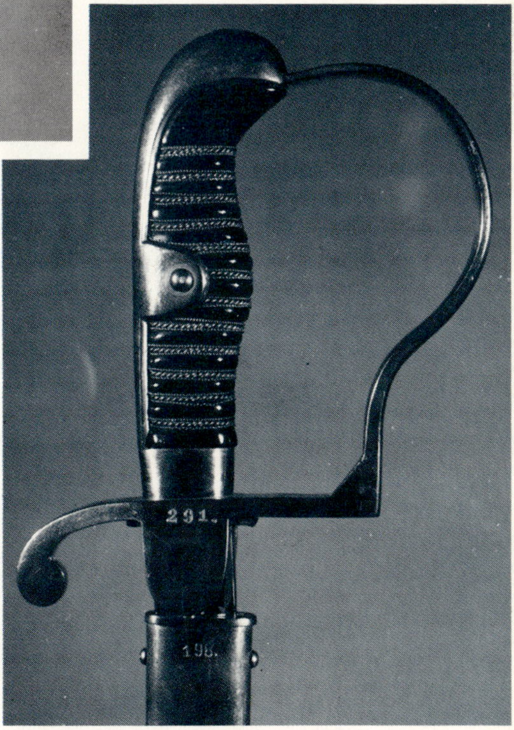

ALTHOUGH THE EXACT INTENT IS NOT KNOWN FOR THIS ALCOSO SABER BEARING THE STATE SEAL OF THURINGEN, IT WAS PROBABLY FOR USE BY A CITY OFFICIAL. THE SERIAL NUMBERS ON THE REVERSE LANGET AND SCABBARD WOULD INDICATE THAT THIS PIECE WAS A CONTROLLED ISSUE ITEM.

"DIE WEHRMACHT", 1938

THE SABER ABOVE, WAS MANUFACTURED BY THE ALCOSO FIRM FOR PRESENTATION TO FIELD MARSHAL VON HINDENBURG BY THE CITY OF SOLINGEN ON THE OCCASION OF HIS 70th BIRTHDAY. NOTE THAT THE TRADEMARK IS ALEXANDER KOPPEL RATHER THAN THE NORMAL ALCOSO TRADEMARK.

Below is another example of a unique pattern sword which was created primarily as an exercise in design and creation. The piece was executed by Professor Woenna of the Solingen Trade School, and exists today in the Deutsches Klingenmuseum in Solingen. The handle is sculpured ivory, while the hilt fittings are hollow cast silver with stones inset. The straight double-edged blade is square-tipped, and bears the etched dedication "Me Fecit Solingen" (Solingen Created Me) with swastikas spacing the words. It is here, in the fuller, that the swastika appears, and nowhere else is it incorporated into the design. The sword is without a scabbard.

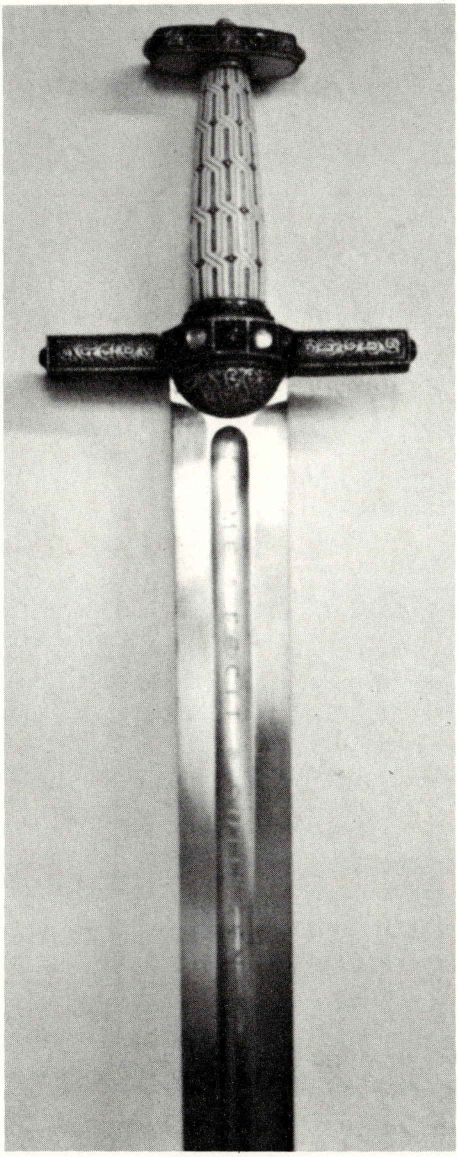

THE COMMANDERS-IN-CHIEF OF THE THREE SERVICES IN THE REICHS CHANCELLRY; GENERAL OBERST VON BRAUCHITSCH (ARMY), GENERALFELDMARSCHALL GÖRING (AIR FORCE) AND GENERAL ADMIRAL RAEDER (NAVY).

The Göring Wedding Sword

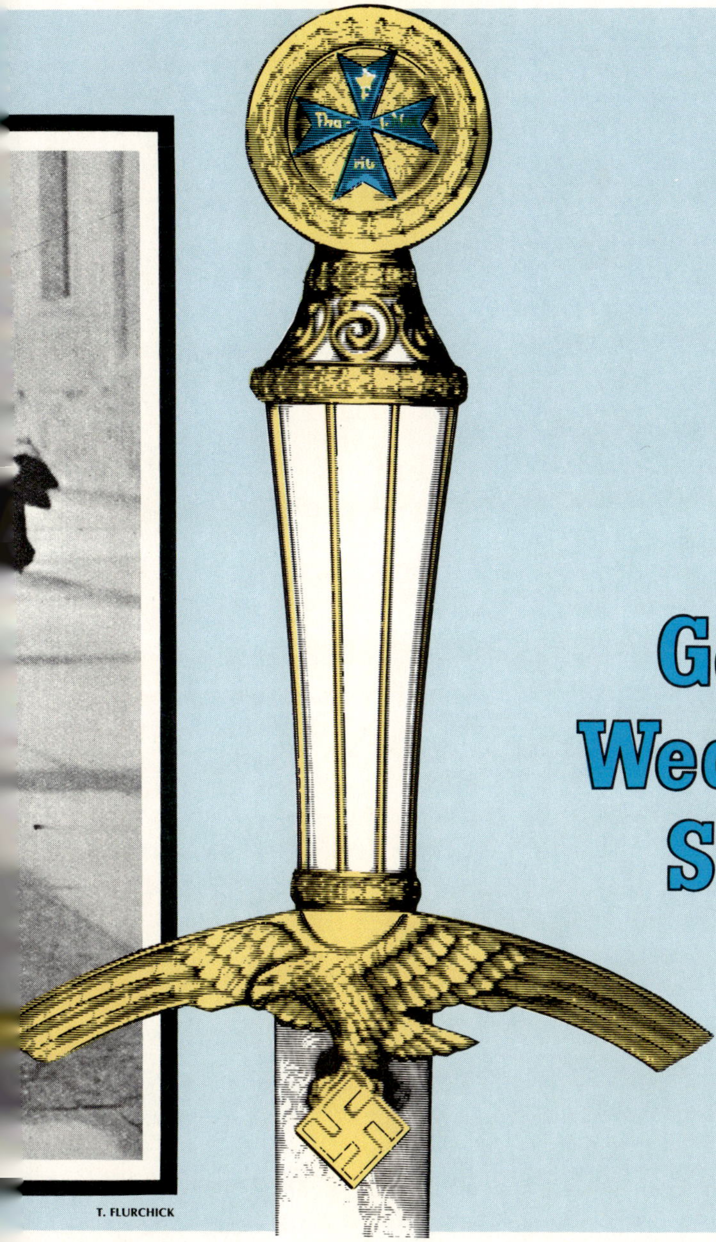

T. FLURCHICK

Only on rare occasions are we able to trace the complete history of a collector's item, and only then if it is an item of historical significance. Where blades are concerned, the best we are able to achieve is to obtain the piece from the veteran who brought it back along with a sketchy account of how it was obtained. In most cases, the story ends there. This is not the case with the famous Göring Wedding Sword, however, since it is a unique blade of historical significance.

When Göring's first wife, Baroness Carin von Kantzow, died, he was grief stricken. With a great deal of pomp and ceremony, Göring had the body of his wife brought from her Swedish homeland and reinterred in Carinhall. He had Carinhall built specifically as a memorial to his beloved Carin around the side of an Imperial hunting lodge on his 100,000 acre estate in Schorfheide (East Prussia). It was here, that Göring spent a great deal of his time and entertained the many visiting state dignitaries during the Nazi diplomatic years.

Carin had been a women of means, and the legacy Göring derived as a result of her death, coupled with the power and prestige gained from his position as President of the Reichstag, made Göring a man of great wealth and importance.

Göring first met Emmy Sonnemann, a beautiful divorcee in her middle thirties, in 1932, and they soon established a close and constant relationship. Emmy was well established as an actress in the State Theater, so the presence of the couple drew great attention wherever they went. It was not until March 1935, that the two announced their wedding plans. Hermann and Emmy were married in a spectacular Nazi state ceremony on April 10, with Hitler serving as witness.

Appreciating the significance of the occasion, the Luftwaffe wished to bestow a wedding gift befitting the man and the occasion. Göring was well known for his love of blades, for it was rare that he was seen without one, in one form or another. Representatives of the Luftwaffe contracted with the Eickhorn firm, one of the most noted blade firms in

THE WEDDING OF HERMANN AND EMMY GÖRING-SONNEMANN AT THE DOM IN BERLIN IN 1935, WAS A GREAT STATE OCCASION. HITLER WAS BEST MAN. HERE THE COUPLE ARE SEEN LEAVING TO THE SOUND OF TRUMPETS FROM THE "GENERAL GÖRING" BAND.

Solingen, for the designing and construction of a sword worthy of Göring. From this commission was to stem one of the most unique and beautiful swords ever produced by European blade craftsmen. A great deal of pains-taking effort went into the design and construction of the piece, and it took months to complete. Construction of the sword was no single enterprise. The ornate damascus blade was hand-forged by the master craftsman, Paul Dingler. The hilt and scabbard mounts, detailed to the point of being a true work of art, were the result of the creative genius of Professor Dluzewski of Berlin. Göring himself had a hand in its final appearance when, upon early inspection, he insisted that the sharkskin be dyed a royal blue to match his Luftwaffe uniform. The sword completed, ranking members of the Luftwaffe presented it to Göring on his wedding day as a gift from the officers and men of the Luftwaffe.

Göring was very impressed with his gift. It reflected his love for beautiful and expensive things as well as mirroring his personal achievements. The pommel bore a facsimile of his W. W. I Pour le Merite award for valor on the obverse, while the reverse bore his coat-of-arms.... a mailed fist clutching a ring. The flying eagle crossguard and the blue sharkskin scabbard were representative of the Luftwaffe, which he commanded. But it was the blade, a swordsmith's work of art, which bore the sentiments of the Luftwaffe. The obverse of the damascus blade bore the raised dedication in gold, "10 April 1935 • Die Reichsluftwaffe Ihrem Oberbefehlshaber" (10 April 1935 • The National Air Force to its Commander-in-Chief), while the reverse dedication read, "Getreu dem Führer für Volk und Reich" (Loyal to the Führer for People and Nation).

LIBRARY OF CONGRESS

(LEFT) GÖRING PROUDLY DISPLAYS HIS WEDDING SWORD TO CHARLES LINDBERG, PRIVATE EMISSARY FROM THE UNITED STATES ON JULY 23, 1936. WHEN NOT BEING WORN, THE SWORD HAD A PLACE OF HONOR NEAR THE DOORWAY IN CARINHALL.
(BELOW) GÖRING IS SHOWN WEARING THE MODIFIED PATTERN WEDDING SWORD WITH ITS DISTINCTIVE CONVEX HANDLE.

Göring wore the sword at the many Nazi functions and ceremonies throughout his career. Its great weight (5lbs, 11ozs) caused him to give way to the wear of his "Fliegergeneralsdegen". The wedding sword was no less his prized blade, and it was afforded a place of honor on the wall at Carinhall. As the tide of war shifted, Göring wished to preserve the great art treasures which he had amassed as a result of his political and military positions. He instructed that his priceless collection of objets d'art be secured in a mountain cave at Königsee, near his Berchtesgaden retreat.

It was here, in May 1945, that the advancing U. S. forces found the treasure trove in "Aladin's Cave". The responsibility for clearing the cave fell to Platoon Sergeant Bob Thibodaux of the 1269th Engineer Combat Battalion. He, and a detail of three men, entered the cave which had seven or eight rooms. Only one room, however, with its entrance sealed with cement, contained any treasure. An effort had been made to camouflage the entrance... attempting to make the sealed entrance simulate the walls of the cave. Had it not been for the German caretaker who led the Americans to the cave, the trove may have remained one of the great lost treasures of the war. The cement, however, had not completely dried, and wet spots could be seen for days after the cave was found. It was because of the dampness that the former caretaker of the treasure led the Americans to its location in the first place. He was concerned that the damp environment of the cave would ruin the priceless treasure and his concern for the irreplaceable art trove weakened his loyalty to Göring. Besides, Göring by this time was already in the hands of the Americans, having surrendered to the U. S. Seventh Army. Sergeant Thibodaux confirms that the art and tapestry portion of the treasure would surely have deteriorated from the dampness had it not been for the caretaker.

U.S. ARMY PHOTOGRAPH

GÖRING, WHO SURRENDERED TO THE U.S. SEVENTH ARMY, ARRIVES FOR DETENTION AT A CAMP ON MAY 9, 1945.

It took the four men working daily for a week to clear the material from the cave. Items such as silverware, swords, table cloths, paintings and tapestries were taken to a central point in the City of Berchtesgaden for disposition. Identifiable items were prepared for shipment back to the owners all over Europe, while the remainder of the artifacts were prepared for later shipment to Berlin. The mission of sorting the treasure was of a secondary nature to Thibodaux's small element as they were primarily concerned with securing the cave from possible GI and German looters. This had been their mission late in the war...the securing of buildings of intelligence interest until such areas were cleared by appropriate intelligence agents. Even to the untrained eye, it was obvious that security of the cave was absolutely necessary. Later estimates by the Seventh Army placed the treasure's value in excess of 169 million dollars.

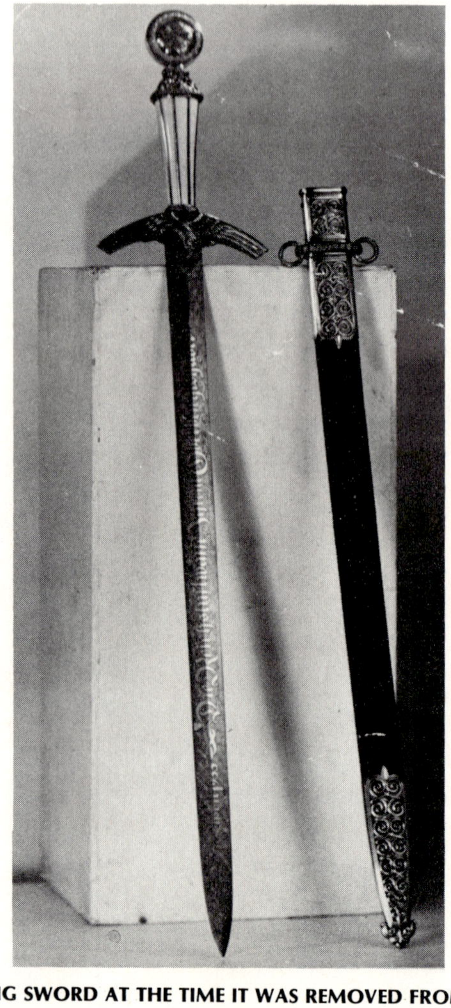

 THE GÖRING WEDDING SWORD AT THE TIME IT WAS REMOVED FROM THE TREASURE CAVE.

PLATOON SERGEANT ROBERT THIBODAUX, 1269th ENGINEER COMBAT BATTALION ATTACHED TO THE U.S. SEVENTH ARMY, SHOWN HOLDING THE GÖRING WEDDING SWORD AFTER ITS REMOVAL FROM THE "ALADIN'S CAVE" AT KÖNIGSEE.

As was the case with selected personal possessions of Göring, Sergeant Thibodaux laid claim to the sword which had been designated the "Ehrengabe der Luftwaffe an Generalfeldmarschall Göring" (Honor gift of the Luftwaffe to General Field-Marshal Göring) by its bestowers. He carried the sword for three months until rumors persisted that his unit was being rotated to the Pacific Theater. He took a chance and prepared the sword for shipment home... no easy task considering its length. Instead of rotating to the Pacific, Sergeant Thibodaux returned to the United States in August 1945, and found the sword had already arrived in undamaged condition as a result of his pains-taking effort to crate the piece. He was relieved that the sword had passed the inspectors, and had not been "liberated" as a war trophy during its shipment home. He proudly displayed the sword in his recreation room until noted collector/author LTC James P. Atwood convinced him to part with the piece. The famous wedding sword changed hands one more time when George R. Canaday, a collector who was visiting LTC Atwood stated after holding the sword, "I have never experienced such a feeling.... I am holding a piece of history". Some time after that, Mr. Canaday acquired the sword. The historical piece currently remains in a local bank vault, but Mr. Canaday has considerately offered to share his experience with serious collectors providing prior arrangements are made.

There is and was only one "Göring Wedding Sword", even though photos will indicate that Göring had at least one copy of the sword made. Why a copy was made is not known, as the Eickhorn firm had explicite instructions from Göring never to strike another sword such as his, and they made no other piece than the original sword. Yet there are discernible differences... a convex grip rather than the concave, a slightly changed pommel mount, and the variation in the upper scabbard fitting.... to indicate that Göring wore two, not one, pattern swords of the same design. One possible explanation is that, due to the excessive weight of the original, he had a copy manufactured from lighter material. Unfortunately, even Frau Göring was unable to shed any light on the matter, but simply stated that Göring treasured the sword commemorating their wedding.

GEORGE R. CANADAY COLLECTION

POMMEL (OBVERSE)

POMMEL (REVERSE)

REVERSE SIDE OF THE BLADE SHOWING THE RAISED GOLD 1935 PATTERN EICKHORN TRADEMARK, THE INITIALS "pd" (PAUL DINGLER, MASTER CRAFTSMAN WHO CREATED THE DAMASCUS BLADE FOR THE EICKHORN FIRM) AND VERY GOOD DETAIL OF THE DAMASCUS PATTERN WHICH DECORATED THE BLADE.

GEORGE R. CANADAY COLLECTION

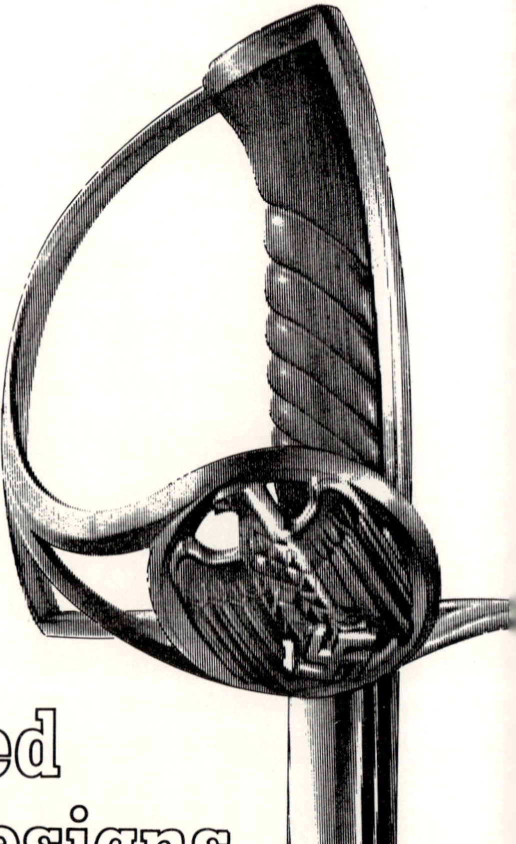

Proposed Sword Designs

The active competition among blade manufacturers to capture the wholesale market caused a great variation of designs to be created. This was especially true of the Army Officer's "Säbel" since no official restriction was imposed limiting the design of the unofficial patterns.

Many designs were created, and most were rejected. A few had not only artistic merit, but the promise of appealing to the consumers' eye, thus eventual consumption. Most designs put forth, originated at the Solingen Trade School by members of the staff and the student blade apprentices. A few firms commissioned professional designers, however, to create potentially lucrative designs for them to consider.

Towards the end of the war, after stringent production limitations had been imposed, prototype swords and daggers were only created in Berlin when constructed of the materials declared as "strategic". This did not prevent the flow of designs, however, in hopes that one or more might be approved. The following three swords are but a few examples of new designs created in 1941, but never placed into production.

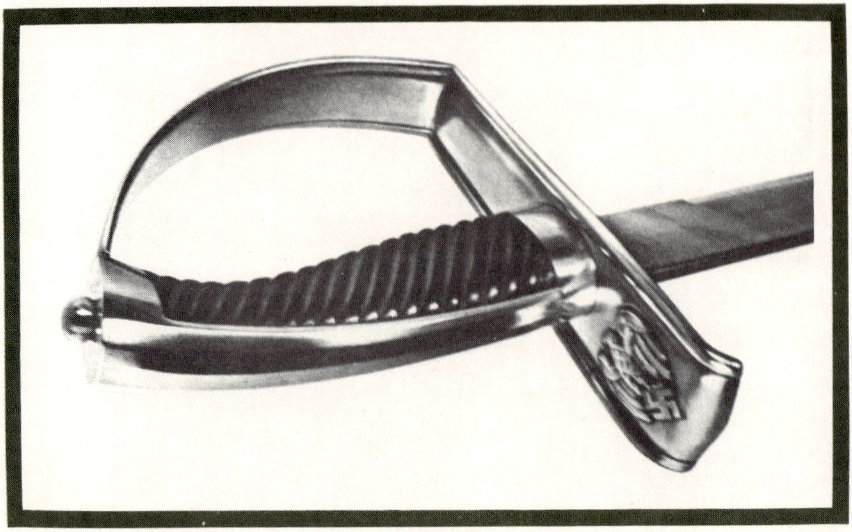

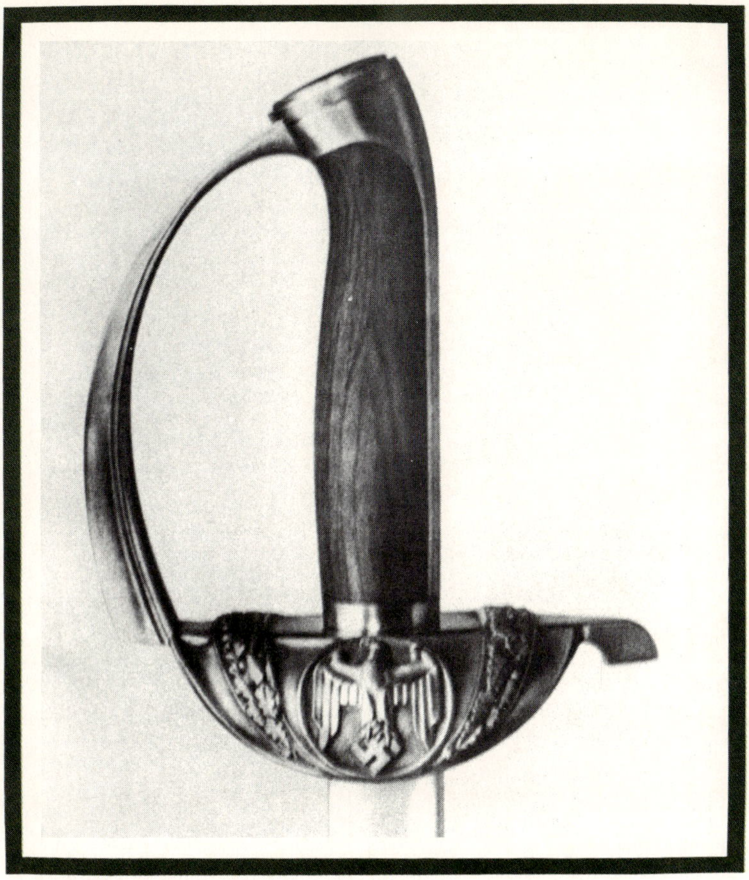

PROPOSED OFFICER'S SWORD (OFFIZIERDEGEN) WITH BOW AND GUARD.
GRIP: ROSEWOOD
METAL FITTINGS: BRONZE
SCABBARD: OXIDIZED MATAL

PROPOSED ARMY OFFICER'S SWORD.
GRIP: ROSEWOOD
GUARD: BRONZE EAGLE ON OXIDIZED FITTINGS
SCABBARD: OXIDIZED MATAL

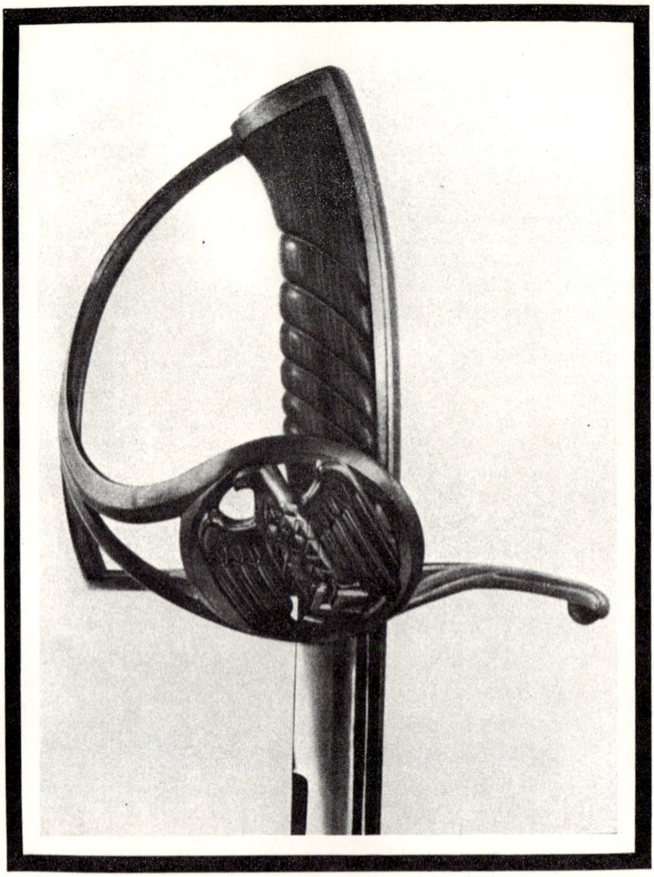

PROPOSED ARMY OFFICER'S SWORD WITH HANDGUARD.
GRIP: ROSEWOOD
METAL FITTINGS: BRONZE
SCABBARD: OXIDIZED METAL

THE PROPOSED FIELD MARSHAL'S SWORD

With great pomp and ceremony, the Supreme Commander of the German Armed Forces, Adolf Hitler, recognized the service performed by his military leaders upon the successful conclusion of military campaigns in Belgium and France. On July 19, 1940, the promotions of one Reichsmarschall of the Greater German Reich and twelve General-Feldmarschälle were placed before the members of the Reichstag. All received unanomous approval. Chief of the Luftwaffe, Hermann Göring, was accorded the singular honor, while the nine Army and three Luftwaffe generals to be honored were:

Walther von Brauchitsch - Commander-in-Chief of the Army
Wilhelm Keitel - Chief of the High Command of the Armed Forces (OKW)

Gerd von Rundstedt	- Commander-in-Chief, Army Group "A"
Fedor von Bock	- Commander-in-Chief, Army Group "B"
Wilhelm von Leeb	- Commander-in-Chief, Army Group "C"
Erwin von Witzleben	- Commander-in-Chief, First Army
Günther von Kluge	- Commander-in-Chief, Fourth Army
Walter von Reichenau	- Commander-in-Chief, Sixth Army
Wilhelm List	- Commander-in-Chief, Twelfth Army
Erhard Milch	- Inspector General of the Luftwaffe
Albert Kesselring	- Chief of the Luftwaffe 2, Northwest
Hugo Sperrle	- Chief of the Luftwaffe 3, South

Each was given his badges of rank, including the Feldmarschall's baton, at the end of August 1940.

V. BRAUCHITSCH

KEITEL

V. RUNDSTEDT

V. BOCK

V. LEEB

V. WITZLEBEN

LIST

MILCH

V. KLUGE

V. REICHENAU

KESSELRING

SPERRLE

Seeking to create a sword commensurate to the rank, the Solingen Chamber for Industry and Commerce designed a sword they hoped would be acceptable. The description which accompanied the designs read as follows:

> "The Field Marshal's sidearm shows early forms of the Germanic sword with straight, double-edged blade. It fulfills its function...to distinguish, in addition to the Marshal's baton, its wearer's special rank by its special design, the rich material and the lavish but subdued embellishments. Because the grip is fashioned of transparent smoked topaz, the grip is strong yet light in weight. The knob of stainless steel shows the Wehrmacht eagle on the front side and the initials of the wearer on the back. The linear ornaments of the grip, the swastika and the initials are applied in gold. On the obverse of the double-edged, wide sword blade runs a blood groove showing the motto, "Alles für Deutschland". The reverse of the blade is flat and provides space for listing battles and merits of the sword wearer. This individualized the weapon, which is bestowed in the same way as the Marshal's baton. The sheath is of calf parchment, a durable German raw material with a very lively pattern."

The military setbacks and the resultant disfavor in which Hitler looked upon his military leaders would possibly account for the reason the proposed design was never placed into production.

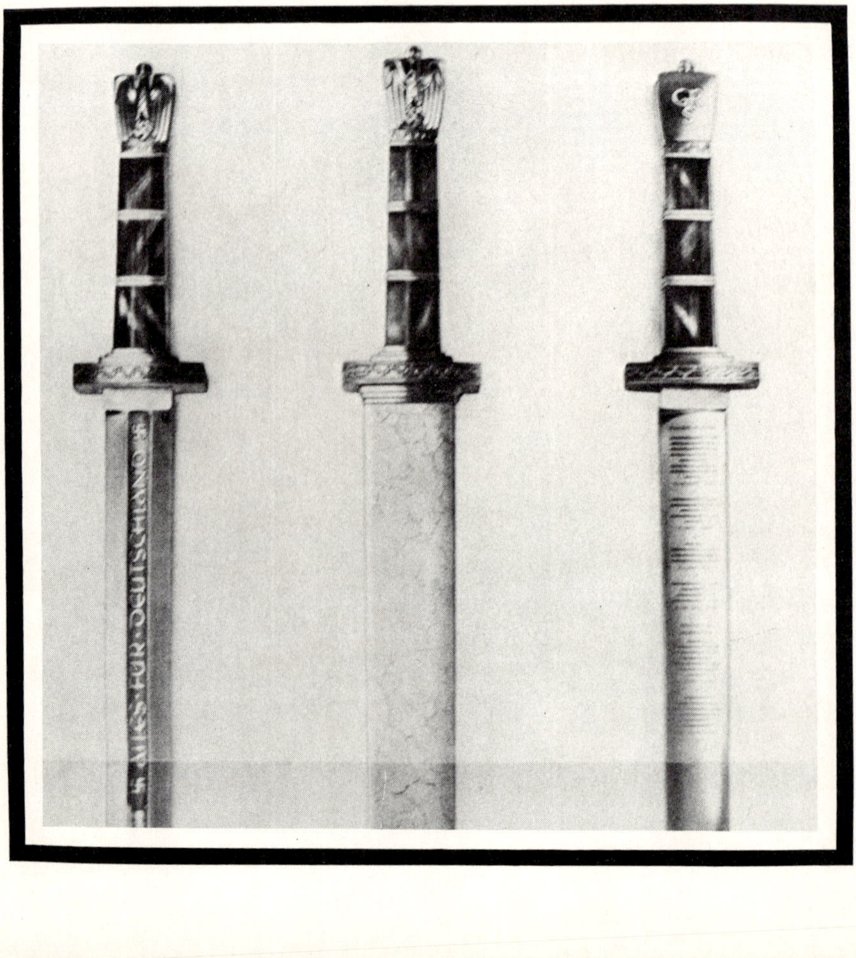

PROPOSED SWORD FOR FIELD MARSHALS OF THE GERMAN REICH, 1941.
METAL PARTS OF GRIP AND SHEATH: STAINLESS STEEL
LINEAR ORNAMENT, SWASTIKA AND MONOGRAM: GOLD
BLADE: HAND-WROUGHT AND ETCHED
SHEATH: CALF PARCHMENT

THREE PRESENTATIONS OF THE FIELD MARSHAL'S SWORD. AT LEFT IS THE OBVERSE OF THE SWORD OUT OF THE SCABBARD. CENTER SHOWS THE SWORD IN ITS CALF PARCHMENT SCABBARD. RIGHT SHOWS THE INITIALS OF THE WEARER ON THE POMMEL REVERSE, AND THE RAISED GOLD ACCOUNTS OF THE SIGNIFICANT BATTLES FOUGHT BY THE WEARER.

GRAND ADMIRAL RAEDER AND GENERAL FIELD MARSHAL BLOMBERG ACCOMPANYING HITLER AT A KEEL-LAYING CEREMONY, AT KIEL.

CHARLES CAMPBELL

Military Daggers

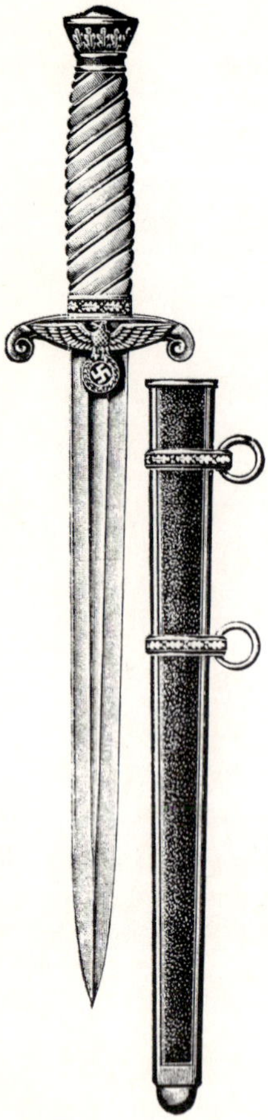

THE ARMY OFFICERS DAGGER

THIS ARMY ENGINEER LIEUTENANT POSES WITH HIS ARMY DAGGER WORN SUSPENDED FROM BENEATH HIS TUNIC.

Luftwaffe Hauptmann Joachim Uebe is shown below wearing the Army Officer's Dagger, representative of his former service arm. He entered military service in December 1936, as a cadet in the 12th Artillery Regiment. In September of the following year, he entered officer training and was commissioned in 1938. As an Army officer, he underwent observer training in 1938. One year later, Göring observed this Army officer taking part in flight observation and asked why he was'nt in the Luftwaffe. By order of Göring, Lt. Uebe was transferred into the Luftwaffe. Even though he was required to change to the Luftwaffe uniform, he continued to wear his Army breeches, boots and dagger in contradiction to existing regulations. When assigned to a combat area, the dagger was relegated to a position on the wall of his bunker. Defiance of regulations such as this might partially account for some of the most unusual uniform combinations observed in numerous photos.

J. EUBE

HAUPTMANN EUBE AND A FRIEND.

HAUPTMANN EUBE'S BUNKER IN 1944 (NOTE ARMY DAGGER ON THE WALL).

SA-Stabschef Viktor Lutze was presented with the Army Honor Dagger on his 50th birthday by Generalfeldmarschall Keitel on behalf of the Oberkommando der Wehrmacht.[1] Lutze is shown below addressing Army troops in Paris in 1941.

The presentation case for the Lutze Army Honor Dagger was lined inside with a rich green velvet and trimmed with a white, silk twisted cord. The wood case is covered with a pebbled black contact paper, and is secured by two brass hinged clasps. The dagger and case were constructed by the Alcoso Firm.

NATIONAL ARCHIVES

---

[1] Uniformen-Markt, January 15, 1941, page 15.

THE NAVY OFFICER'S DAGGER

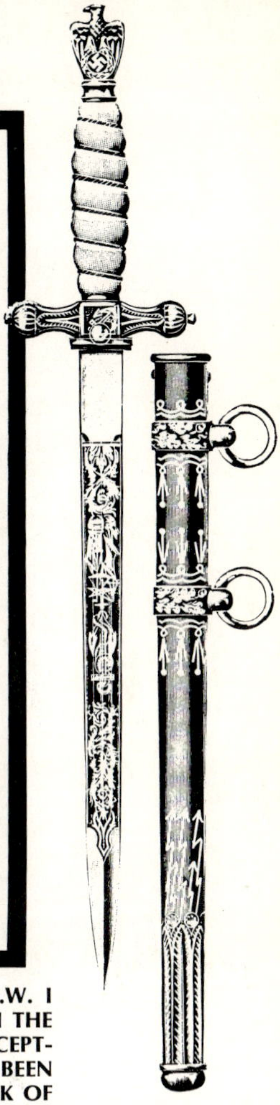

SHOWN HERE IS GRAND ADMIRAL RAEDER WEARING A W.W. I PATTERN NAVY DAGGER WHICH HAS BEEN MODIFIED WITH THE WEHRMACHT EAGLE POMMEL. THIS MODIFICATION WAS ACCEPTABLE BY REGULATIONS PROVIDING PRIOR SERVICE HAD BEEN HONORABLE, AND THE CHANGE DID NOT AFFECT THE RANK OF THE WEARER.

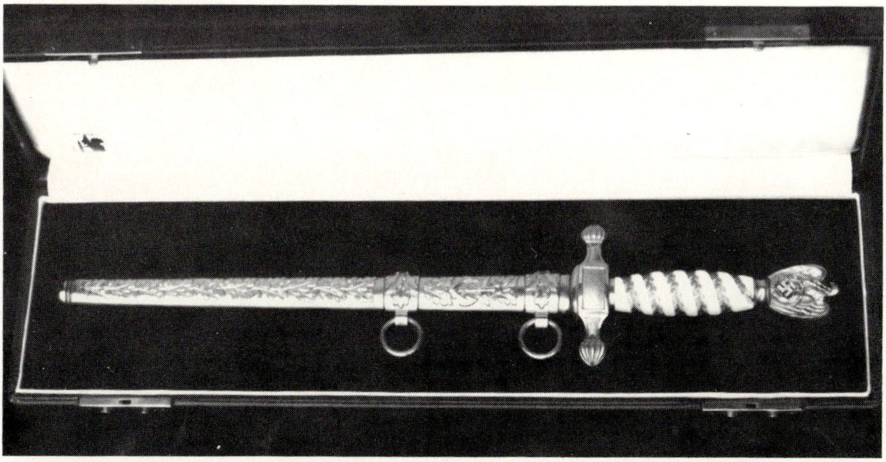

WAGNER-PIAR COLLECTION

NAVY HONOR DAGGER (EHRENDOLCH) AWARDED TO GENERAL ADMIRAL CARL WITZELL ON SEPTEMBER 30, 1942, SHOWN HERE IN THE ORIGINAL PRESENTATION CASE. WITZELL WAS THE THIRD RECIPIENT OF THE RAEDER NAVY HONOR DAGGER. ALL SPECIMENS WERE PRODUCED BY THE EICKHORN FIRM.

Illustrated below is a presentation case for the Navy Honor Dagger presented by Grand Admiral Karl Dönitz. Since the dagger was only manufactured by the Eickhorn firm, the case bears the Eickhorn trademark as well. Shown here is the late pattern trademark, which corresponds with the period of manufacture. Unlike the cases which contained the Raeder Navy Honor Daggers, the lid of the Dönitz case bears the Navy eagle in gold, whereas the Raeder lid has a gold, fouled anchor in the lower right corner. The inside top of the case is lined with white satin with the trademark of the Eickhorn firm printed in gold (1941 pattern lower left side for the Raeder case, and 1942 pattern lower right side for the Dönitz case). The bottom is lined in blue velvet and is recessed and contoured to accept the dagger. The dark blue leather covered case measures 18" x 5" x 2", and is secured by two brass slide clasps.

PHOTO: ANDY WALKER

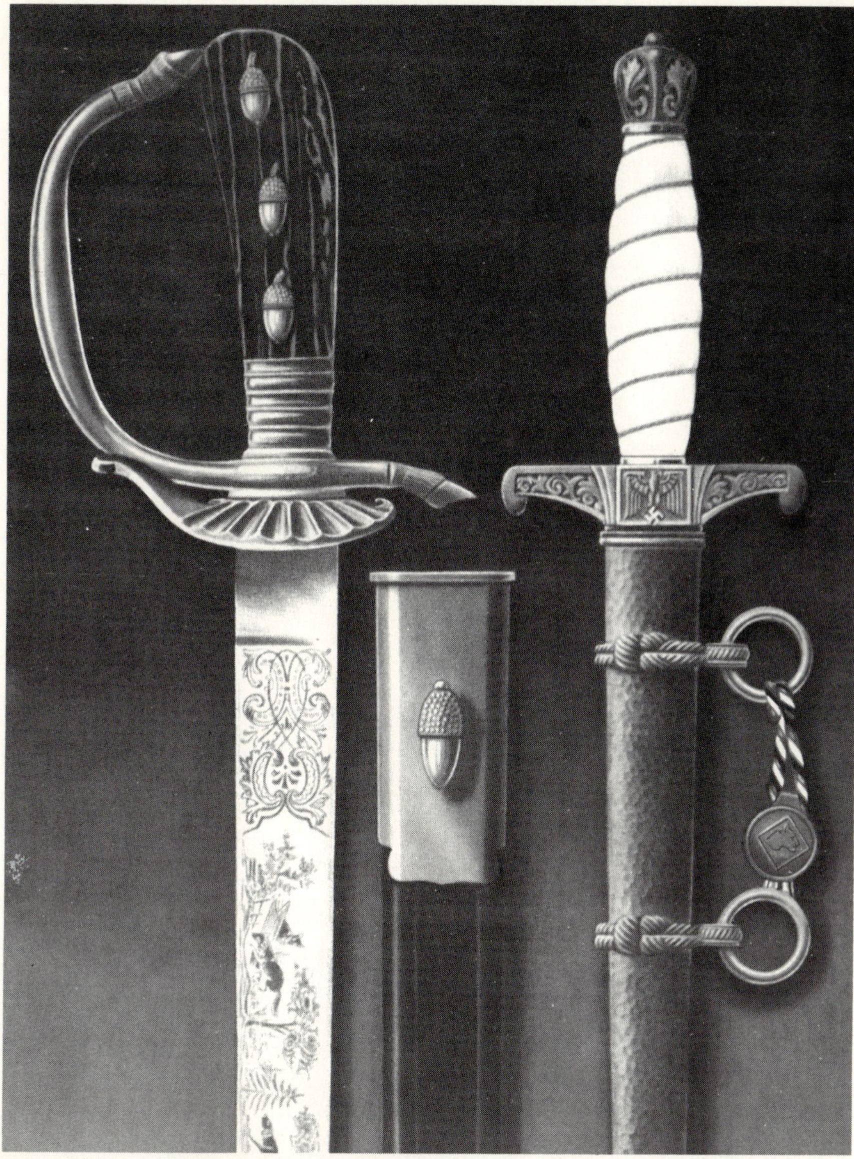

PUMA-WERK

THE ABOVE RENDERING IS FROM A 1940 PUMA SALES CATALOGUE WHICH ILLUSTRATES (LEFT) NO. 10 GERMAN FORESTRY OFFICIAL'S KNIFE, AND (RIGHT) NO. 8 GERMAN NAVAL OFFICER'S DAGGER.

The Puma firm has gratiously permitted me to take this illustration from their sole remaining copy of their 1940 sales catalogue. It clearly shows that such a pattern Navy dagger did, in fact, exist and was the sole production design of the Puma-Werk. Designated Model Nr. 8 Deutscher Marineoffizierdolch, it is the standard pattern Navy sidearm and was not intended for specified use, as alleged, by naval assault troops. It is interesting to note that this is one of three authorized Navy daggers available in 1940. The other two were the basic pattern which is normally found with the Wehrmacht eagle pommel, and the other was a production pattern of the Alcoso firm having the "onion-ball" pommel. One modification of a Navy dagger which is still without recorded purpose is a piece produced by WKC with a black celluloid grip over wood, with the standard wire wrap. Type of manufacture and date of the blade trademark gives every indication that the piece was produced after 1938, and is not made up from a grip belonging to a 1919 pattern having a solid black grip.

As the war progressed, material became more scarce and production techniques were varied to make use of all available materials. For example, specimens have been found constructed with solid steel scabbards with fixed (welded) swivel rings, and crossguards which do not have the scabbard release button, but with a blade which is grooved for the retaining spring.

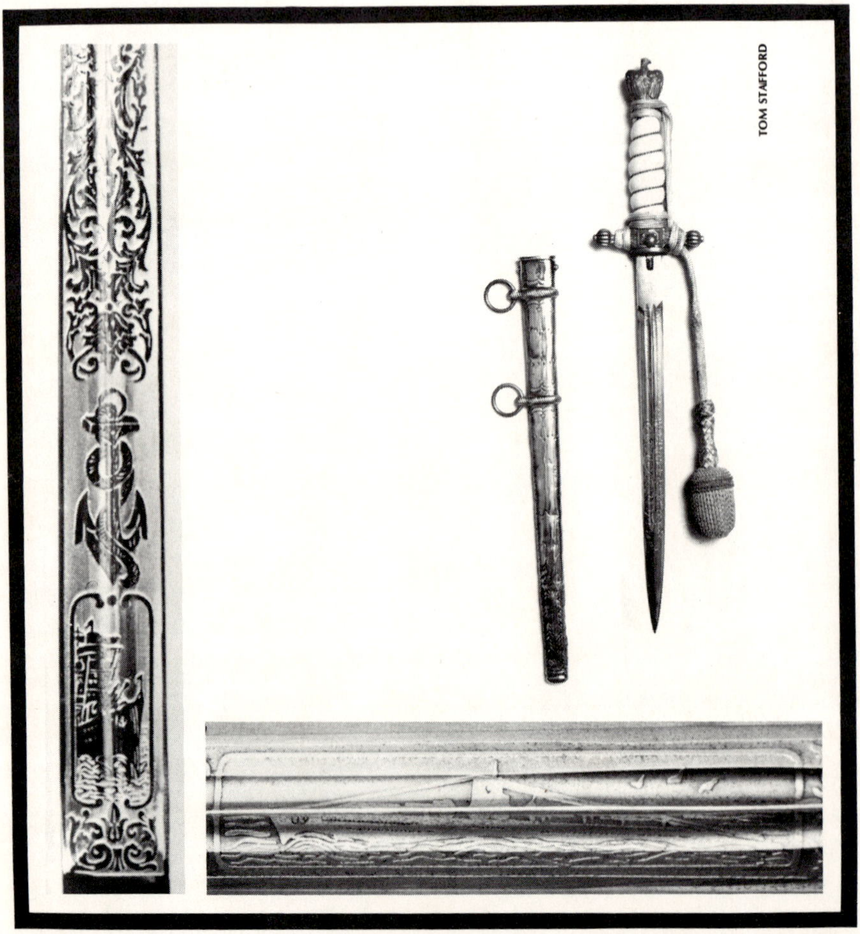

THIS SPECIAL, ETCHED PATTERN BLADE WAS PRODUCED BY THE KLASS FIRM AND SHOWS A CRUISER AND FOULED ANCHOR ON THE OBVERSE, AND A SUBMARINE BEARING THE MARKINGS "U. 9" ON THE REVERSE.

PHOTO: ANDY WALKER

SPECIAL NAVY PRESENTATION MANUFACTURED BY WKC SHOWING THE REGULATION WRAP FOR THE NAVY PORTEPEE. THE SCABBARD TIP HAS THE OWNER'S NAME ENGRAVED IN SCRIPT. NOTE THAT THE HANGER RINGS ARE ENGRAVED. IN KEEPING WITH THE PURCHASE OF A HIGH QUALITY DAGGER, THIS SPECIMEN ALSO HAS THE GENUINE IVORY GRIP.

"DIE WEHRMACHT", 1938

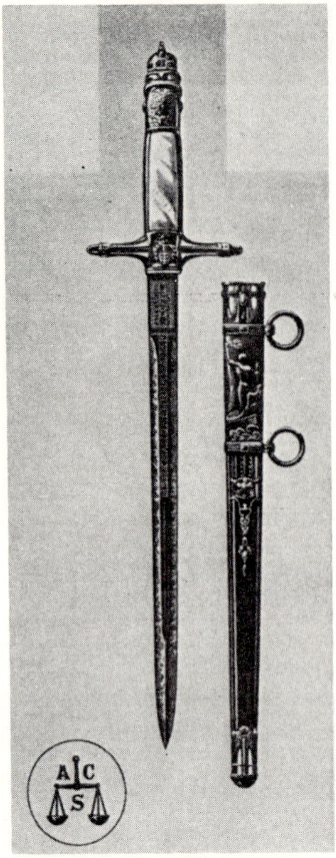

A SPECIAL NAVY HONOR DAGGER PRODUCED BY THE ALCOSO FIRM FOR ADMIRALS VON SCHEER AND VON HIPPER.

PHOTO: FRED STEPHENS

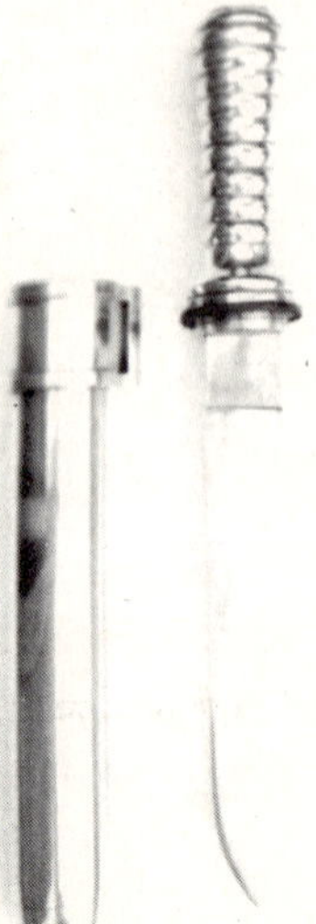

GERMAN FROGMEN SYNCRONIZE THEIR WATCHES BEFORE DEPARTING ON AN UNDERWATER MISSION. NOTE THAT THE UTILITY KNIFE WORN SUSPENDED UPSIDE DOWN, AT THE SIDE, IS SIMILAR TO THAT WORN BY THE ITALIAN FASCIST MVSN TROOPS.

"MÜNCHENER ILLUSTRIERTE PRESSE", 1944

(a) A GERMAN NAVY DEEP-SEA DIVER PREPARES TO DESCENT TO THE DEPTHS. HIS DIVING UTILITY KNIFE CAN BE SEEN JUST BELOW HIS ELBOW, ENCASED IN ITS METAL SCABBARD. (b) A CLOSE-UP OF THE DIVERS UTILITY KNIFE. AN ALL METAL CONFIGURATION, THE KNIFE PORTION SCREWS INTO THE SCABBARD TO PROTECT THE BLADE FROM RUST AND LOSE. BOTH PIECES ARE STAMPED WITH THE NAZI ORDNANCE PROOF.

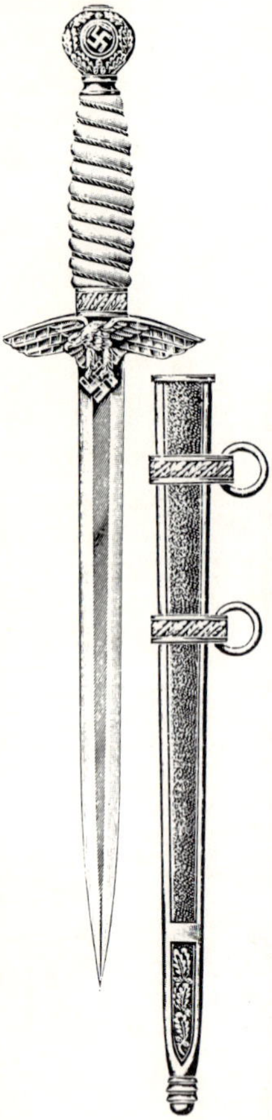

THE AIR FORCE OFFICER'S DAGGER

General Wever, the first Chief-of-Staff of the newly formed Luftwaffe, is shown above wearing the "Luftsport-Verband" Flyer's Dagger. General Wever was killed in an air crash on June 3, 1936, and to commemorate his name, Bomber Wing 4 took the designation, "Geschwader General Wever". This pattern dagger was the forerunner of the similar appearing dagger adopted for use by the Luftwaffe, upon its formation in 1935.

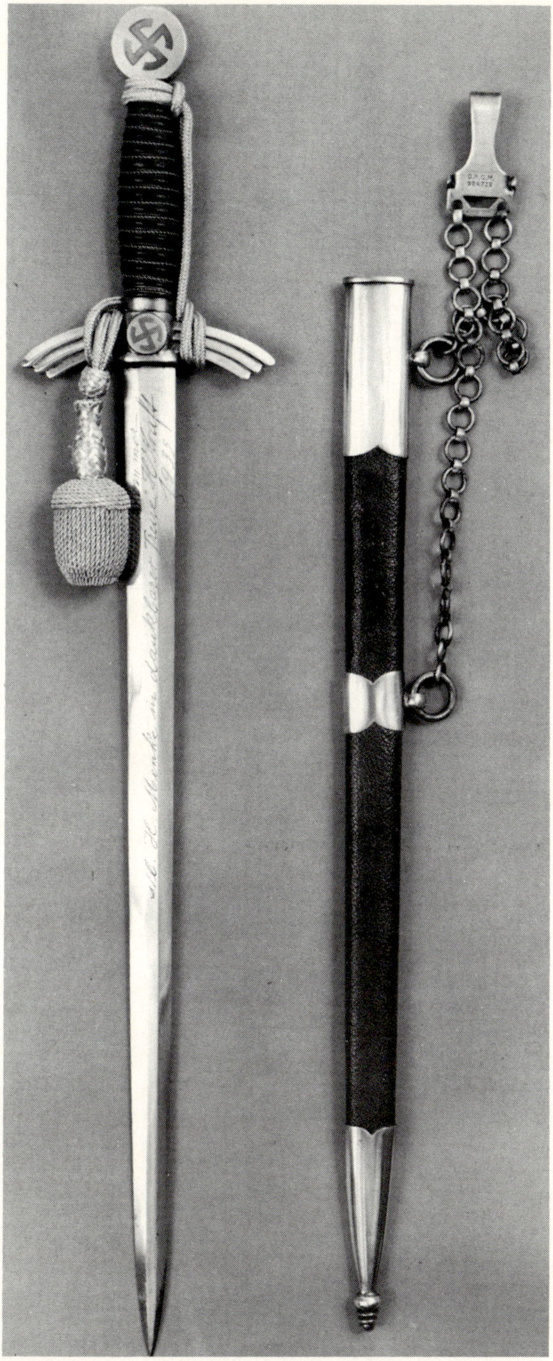

MAJ. TOM JOHNSON

THE STANDARD PATTERN "LUFTSPORT-VERBAND" FLYER'S DAGGER WHICH WAS INTRODUCED PRIOR TO THE COMMONLY KNOWN "FLYER'S KNIFE". THIS SPECIMEN HAS A DEDICATION ENGRAVED ON THE BLADE AND DATED 1935. THE DATE, ONE YEAR AFTER THE INTRODUCTION OF THE "DLV FLIEGERMESSER" (FLYER'S KNIFE), WOULD INDICATE THAT THE EARLY PATTERN "LUFTSPORT-VERBAND" FLYER'S DAGGER WAS WORN WELL AFTER THE INTRODUCTION OF THE KNIFE, WHICH WAS DESIGNATED TO REPLACE IT. THIS PATTERN DAGGER WAS WORN BY ALL RANKS SERVING ACTIVELY WITH THE DLV.

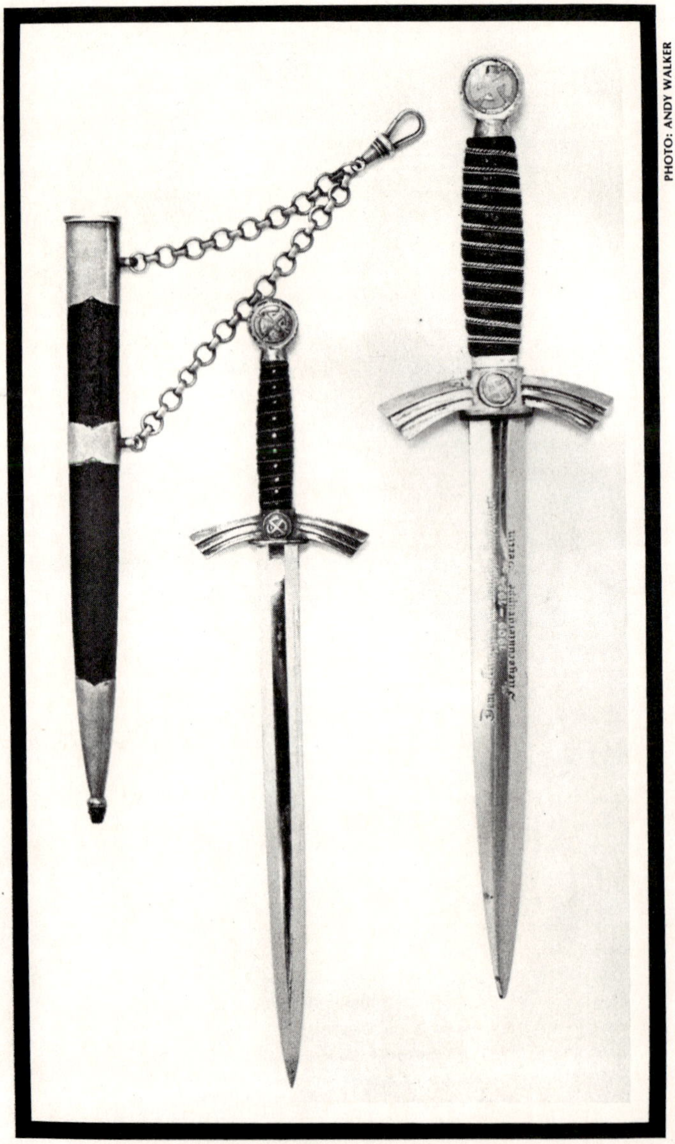

PHOTO: ANDY WALKER

AT LEFT IS A 1934 PATTERN MINIATURE LUFTWAFFE DAGGER, COMPLETE WITH SUSPENSION CHAIN. AT RIGHT IS A HIGH QUALITY PRESENTATION MINIATURE APPROXIMATELY ONE-HALF THE FULL-SCALE VERSION. THE SWASTIKAS IN THE POMMEL AND CROSS-GUARD ARE INLAID WITH AN ORANGE-COLORED ENAMEL. THE BLADE INSCRIPTION READS, "DEM KÄMPFER FÜR DEUTSCHE LUFTFAHRT 1909-1934 FLIEGERUNTERGRUPPE BERLIN".

A LUFTWAFFE OFFICER TAKES A BREAK AT A POLITICAL RALLY. NOTE THAT THE 1937 PATTERN LUFTWAFFE OFFICER'S DAGGER IS WORN SUSPENDED FROM A BELT ATTACHMENT WORN UNDER THE TUNIC WHEN NO OUTER BELT IS WORN.

JOHN SCHLEICHER

CRIS FARLOWE

GENERALFELDMARSCHALL HERMANN GÖRING WEARS THE STANDARD PATTERN 1937 LUFTWAFFE OFFICER'S DAGGER, SUSPENDED FROM DOUBLE STRAP HANGERS. A SILVER PORTEPEE MEASURING 23cm WAS STANDARD FOR ALL OFFICERS, AND AUTHORIZED FOR WEAR BY NCO'S ON FLYING STATUS OR DESIGNATED "UNTEROFFIZIER MIT PORTEPEE".

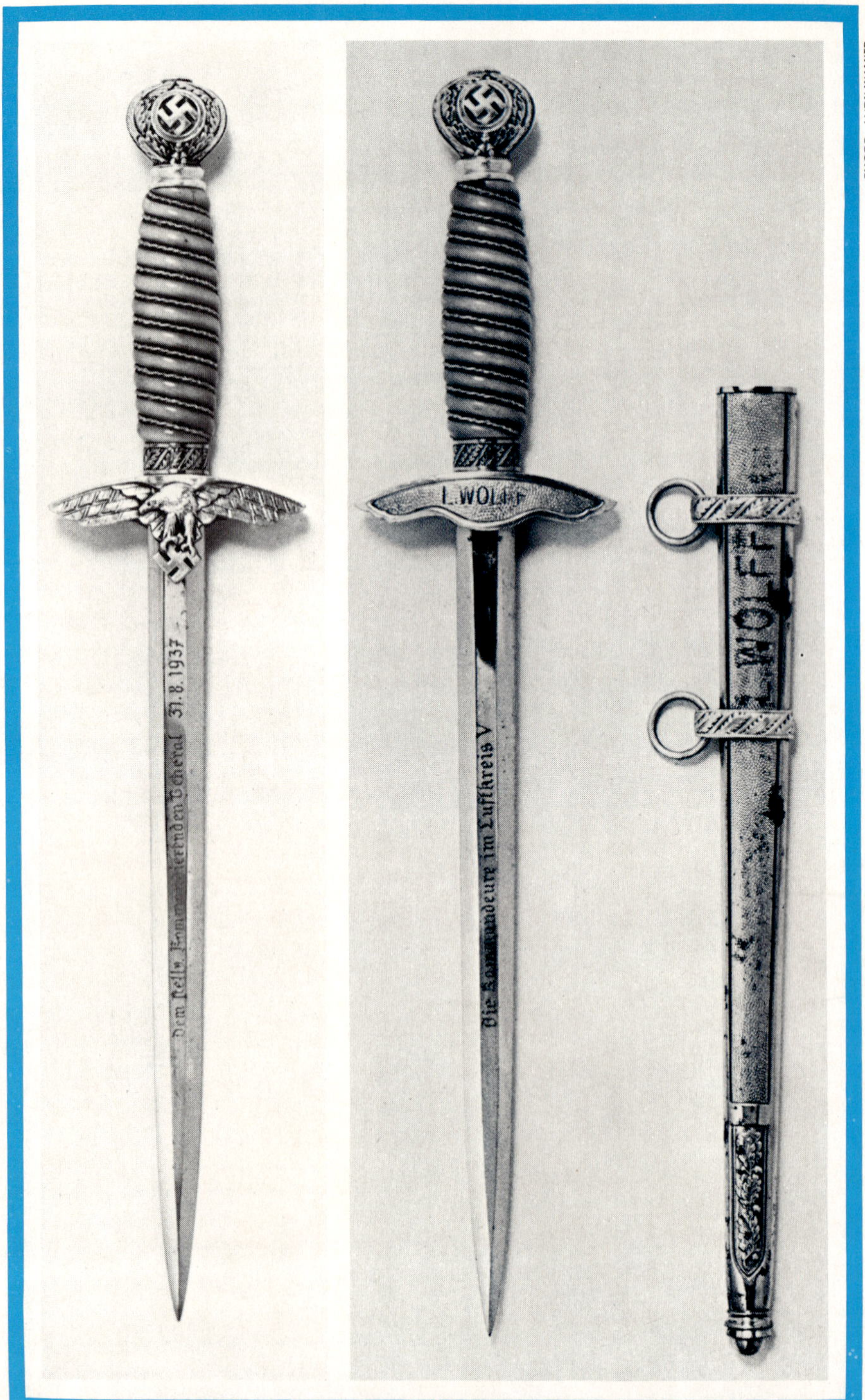

PHOTO: ANDY WALKER

MODEL 1937 PATTERN LUFTWAFFE DAGGER BELONGING TO GENERAL LUDWIG WOLFF, LATER COMMANDING GENERAL OF LUFTGAU XI. ENGRAVED DEDICATION ON THE BLADE READS, (OBVERSE) "DEM STELLV. KOMMANDIERENDEN GENERAL 31.8. 1937", (REVERSE) "DIE KOMMANDEURE IM LUFTKREIS V". MANUFACTURED BY SMF.

PHOTO: ANDY WALKER

MODEL 1937 LUFTWAFFE DAGGER WITH SIMULATED DAMASCUS BLADE BEARING THE RAISED INSCRIPTION, "O. WACHTMEISTER GEORG LUCKAS FLAK. ABTL. 29". MANUFACTURED BY PUMA. GRIP COLOR IS ORANGE.

SA-STABSCHEF LUTZE, ALONG WITH GAULEITER WAGNER AND OTHER MEMBERS OF THE PARTY, PATIENTLY AWAIT THE ARRIVAL OF THE TRAIN BEARING GENERAL RUSSO, ON HIS VISIT TO MUNICH FROM ITALY IN JULY 1938. LUTZE WEARS HIS GOLD FELDHERRNHALLE DAGGER SUSPENDED FROM HIS LEATHER BELT. RUSSO WAS PRESENTED ONE WITH A WHITE GRIP DURING THIS TRIP.

LIBRARY OF CONGRESS

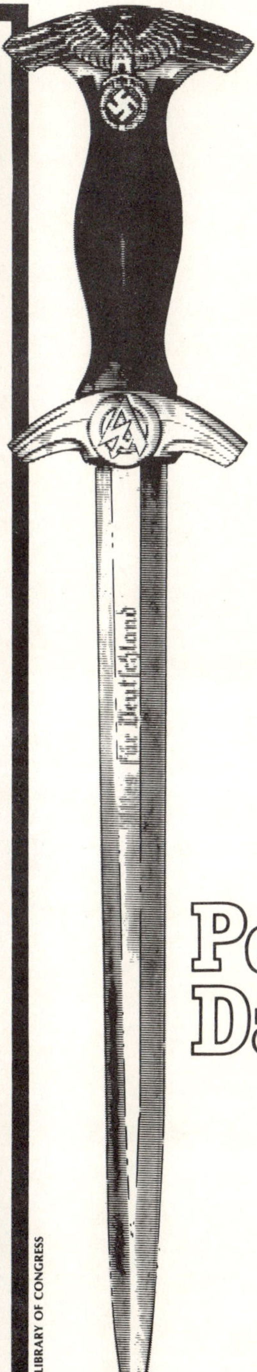

Political Daggers

THE SA

This SA/SS Honor Dagger was worn exclusively by SA-Stabschef Ernst Röhm, who simultaneously commanded the SA and SS during his tenure. His Honor Dagger was possibly the forerunner of those which were authorized after his murder in 1934. Note the unique link suspension hanger. It is probable that this dagger was destroyed immediately after his purge.

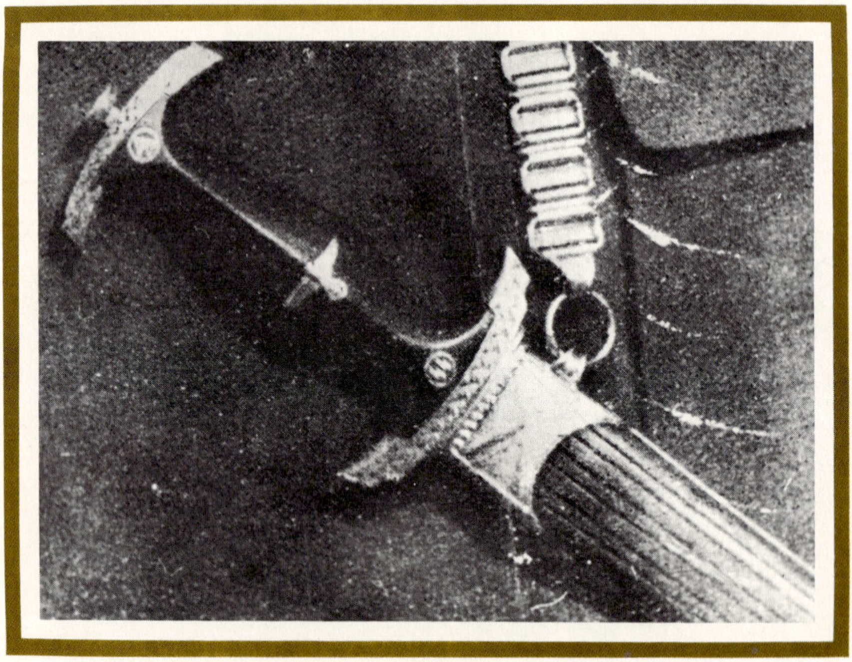

ERNST ROHM'S SA/SS HONOR DAGGER.

Illustrated at right is a 1935 Pattern SA Honor Dagger of the SA-Stabschef, Viktor Lutze, with damascus blade. As can be observed by the 1941 Eickhorn trademark, however, this pattern dagger continued to be bestowed by Lutze long after the adoption of the chained version of the SA Honor Dagger. It is the opinion of this author that the SA, SS and NSKK Honor Daggers were the exclusive production of the Eickhorn firm.

PHOTO: ANDY WALKER

THE SA-GRUPPENFÜHRER BELOW WEARS AN SA HONOR DAGGER.

(REVERSE)

NATIONAL ARCHIVES

Note the SA Service Dagger with chain suspension at right, but unlike any that are of the official pattern. The manufacture date of the dagger is pre-1936. No other information is available on this pattern dagger at this time. The photo shows an SA-Obergruppenführer wearing a similar dagger, but with longer chain suspension, and the lack of any grip inlays. He is talking to General von Brauchitch.

SA-OBERGRUPPENFÜHRER BRUCKNER, SA ADJUTANT TO THE FÜHRER, WEARS THE SA HONOR DAGGER WITH DOUBLE STRAP DETACHABLE HANGERS WHILE ATTENDING THE OPENING OF THE AUTOMOBILE SHOW, FEBRUARY 18, 1938.

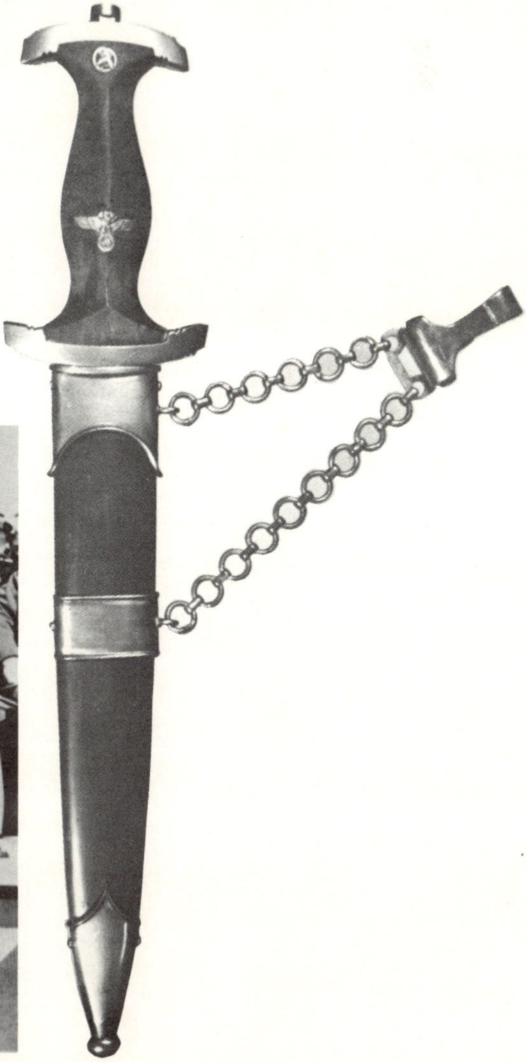

FRED FALLGREN

STANDARD 1933 PATTERN SA SERVICE DAGGER WITH MODIFIED GRIP INLAY. THE METAL SA RUNIC INLAY HAS BEEN REPLACED WITH ONE MADE OF IVORY, AND BEARS THE ADDITIONAL UNIT DESIGNATION, "SA GRUPPE WESTFALEN".

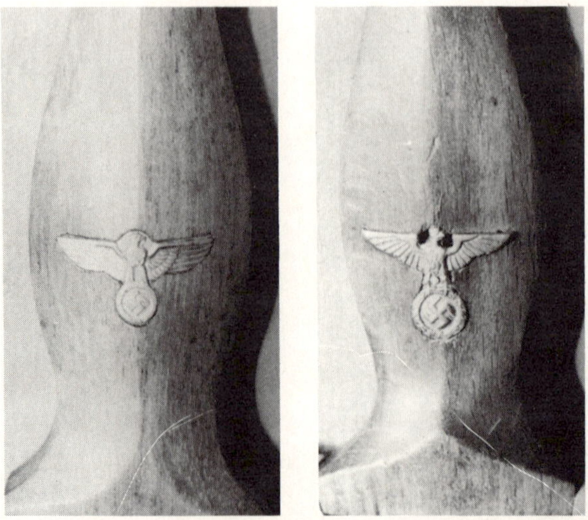

THE SA GRIP TO THE RIGHT IS THE VERY EARLY PATTERN WITH THE ELONGATED NATIONAL EMBLEM WITH POINTED WINGS. THE NATIONAL EMBLEM WAS SILVER-PLATED AND INLAID INTO THE DYED NATURAL-GRAIN WOOD. THIS PATTERN LASTED ONLY A SHORT PERIOD BEFORE STRICT PRODUCTION CODES WERE INTRODUCED. THE PATTERN SHOWN IN THE GRIP AT THE LEFT REPLACED IT, AND BECAME THE STANDARD DESIGN.

SA-FELDHERRNHALLE

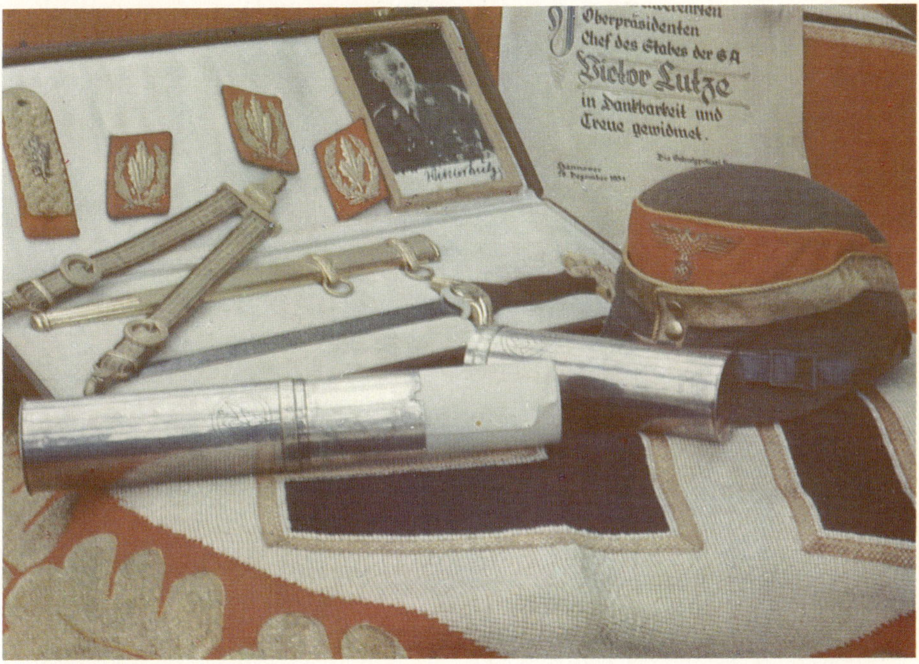

ERIC CAMPION COLLECTION

THE CASED GOLD PRESENTATION SA FELDHERRNHALLE DAGGER WITH DAMASCUS BLADE IS SHOWN ALONG WITH OTHER EFFECTS BELONGING TO SA-STABSCHEF VIKTOR LUTZE, BUT RETURNED TO ENGLAND AS WAR TROPHIES. THE HANGERS WITH GOLD FITTINGS WERE ALSO CASED.

THE SWEARING-IN OF A STURMBANN OF STANDARTE "FELDHERRNHALLE" IN DRESDEN.

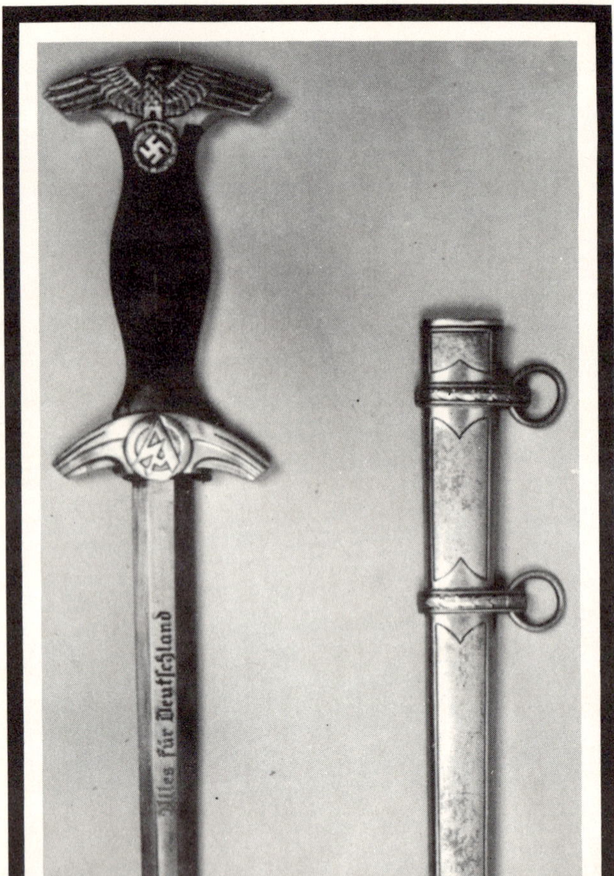

BRUCE REGNEMER

SA FELDHERRNHALLE LEADER'S DAGGER. FITTINGS WERE FINISHED IN SILVER OR GOLD, DEPENDING ON THE GRADE OF THE WEARER. SPECIMENS HAVE BEEN NOTED WITH THE BROWN GRIP, AS SHOWN HERE, AND A WHITE GRIP. THE HANGER FITTINGS BEAR THE SAME METAL COLOR FINISH AS THE DAGGER.

THE SS

The bridal pair stood facing a short column crowned by a basin in which an eternal flame was lit to symbolize the fire of the hearth. They then exchanged rings, as well as bread and salt, the symbols of the soil's fruitfulness and of purity; lastly, the husband gave the wife his dagger in a token of his "Wehrhaftigkeit" (capability of bearing arms) and received another one from his superior SS officer in replacement."[1] This is but one instance of the ceremonial importance placed on the dagger sidearm carried by the SS man. Considering the cost of the dagger, it is doubtful that this ceremony was repeated very often. Every aspect of the dagger was, however, connected with some manner of ritual.

The SS sidearm was considered a weapon of honor, and symbolic of the SS man's ability to defend his own honor and that of the SS and the State. His very entry into the SS was not recognized until the night of November 9 of each year, when SS candidates were initiated into the SS with the bestowal of the SS dagger.

Regardless of his rank, all SS members were designated SS-Mann. Therefore, the dagger sidearm, which was commonly referred to as a man's dagger, was not symbolic of rank but of membership.

Regulations prescribed that the SS dagger be worn only with the service and abbreviated service uniforms.[2] In the case of the service uniform, the SS Service Dagger was worn suspended from the outer leather belt by means of the short leather suspension strap or double chain hanger (from January 1, 1935, to the introduction of the Model 1936 SS Service Dagger, the sidearm was required to be worn suspended from the belt by means of a vertical holder). When the abbreviated service uniform was worn, the dagger was worn attached to a hook which was suspended from under the left pocket flap of the tunic or overcoat. The SS sword was only authorized as a replacement for the SS Service Dagger when the service uniform was worn.

During the war, Reichsführer-SS Himmler ceased awarding the SS Honor Sword, but prescribed that the SS Service Dagger could be worn in its place. By 1944, however, wear of the dagger was supplanted by the pistol sidearm.

---

[1] Berliner Tageblatt, October 11, 1935.

[2] Organizationsbuch der NSDAP, 1935-1943 editions.

The early pattern SS Honor Dagger below, with damascus blade and manufactured by the Eickhorn firm, has some very interesting features. Although not discernible in the photo at left, the reverse of the blade has had a dedication removed. This fact, plus the date of manufacture, would indicate a possible Röhm presentation Honor Dagger. The exact date of this pattern honor crossguard has not been definitely fixed, but it is believed to have been immediately after the removel of Röhm. This would further indicate that the original owner was possibly a recipient of a Röhm dedication blade, had the dedication removed as prescribed by Order Nr. 6747/34, and was subsequently awarded the honor crossguards.

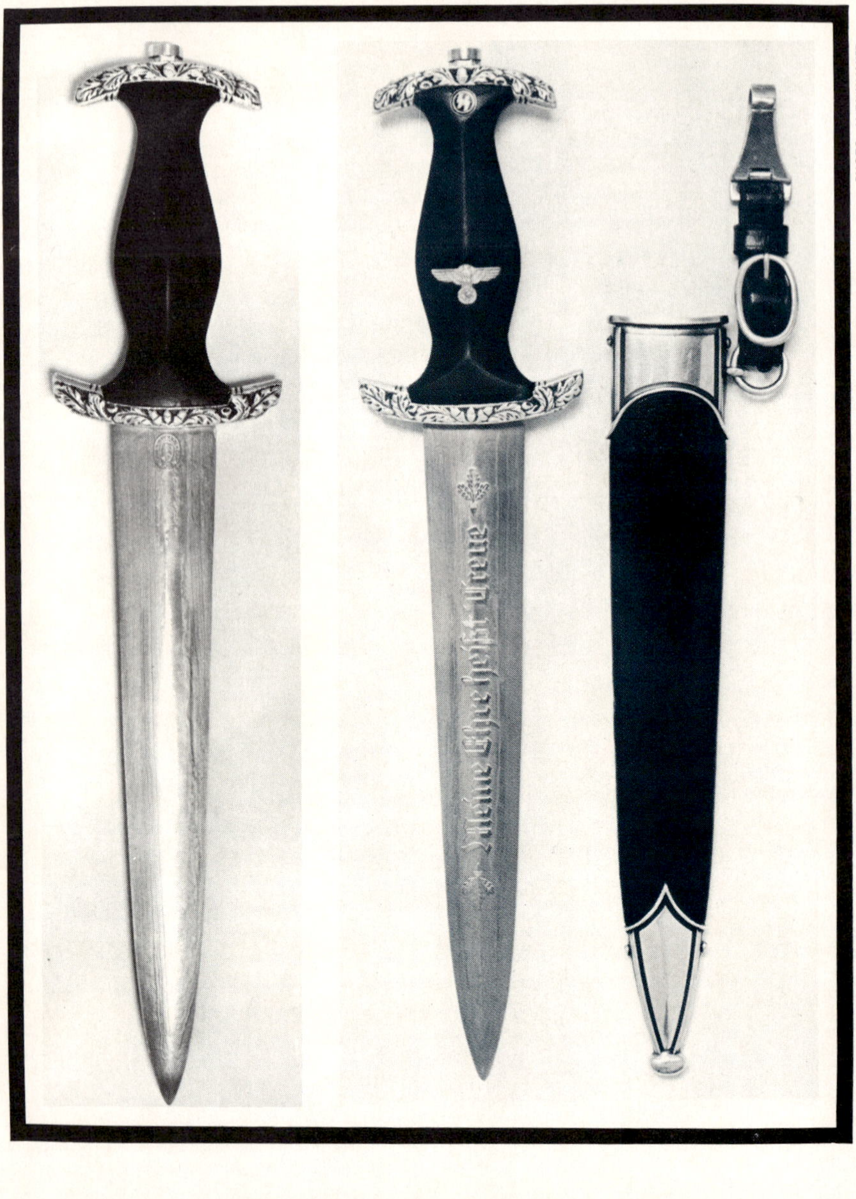

PHOTO: ANDY WALKER

THE NSKK

WAGNER-PIAR COLLECTION

NSKK HIGH LEADER'S HONOR DAGGER. THIS IS IDENTICAL TO THE SA HIGH LEADER'S HONOR DAGGER EXCEPT THAT THE METAL FITTINGS ARE GILT FINISHED, AND THE SCABBARD LEATHER IS BLACK RATHER THAN BROWN. THE ILLUSTRATED SPECIMEN WAS PRODUCED BY THE EICKHORN FIRM.

LIBRARY OF CONGRESS

SA-OBERGRUPPENFÜHRER KUHME CONVERSES WITH NSKK-OBERGRUPPENFÜH EDUARD, PRINCE OF COBURG. THE PRINCE WEARS THE MODEL 1936 SERVICE DAG HAVING ONLY THREE UPPER LINKS IN THE CHAIN PATTERN.

AN NSKK GRUPPENFÜHRER WEARS THE STANDARD PATTERN 1936 SERVICE DAGGER WHILE ATTENDING A COMMEMORATION CEREMONY ATTENDED BY NUMEROUS GAULEITERS, NOVEMBER 1937. NOTE THE FOUR LINKS IN THE UPPER CHAIN.

Since both the SA and SS had a specially designated Honor Dagger, it was only a matter of time before an Honor Dagger for the NSKK became available. The dagger proper, identical to the SA Honor Dagger, was also fitted with the damascus blade. The date of the trademark would indicate that the Honor Dagger was authorized before the actual release of the 1936 Chained Service Dagger. The specimen below bears the signature of NSKK-Korpsführer Hühnlein.

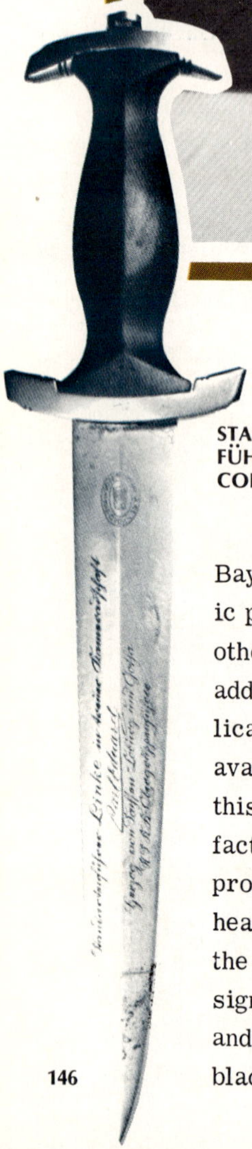

STANDARD 1933 PATTERN NSKK SERVICE DAGGER PRESENTED TO NSKK-GRUPPENFÜHRER LINKE BY NSKK-OBERGRUPPENFÜHRER CARL EDUARD, PRINCE OF COBURG.

The photograph at above right was published in "Daggers, Bayonets and Fighting Knives of Hitler's Germany" to provide graphic proof that a unique pattern High Leader's Dagger did exist. No other information was available at the time of publication, and no additional information has become available. As a result of publication, however, the dagger shown illustrated at right was made available to a dealer, Mr. Ron Fabian, by a veteran who "liberated" this and one other relatively rare dagger. The dagger was manufactured by the Eickhorn firm, who confirms that they were the sole producers of such a pattern dagger. The black grip is set off by heavily silver-plated fittings...the upper guard being identical to the honor pattern, and the lower guard in the form of the NSKK insigne. The scabbard is pebbled and finished with a black burnish, and is quite similar in design to the TeNo Leader's Dagger. The blade is without an organizational motto or inscription.

NATIONAL ARCHIVES

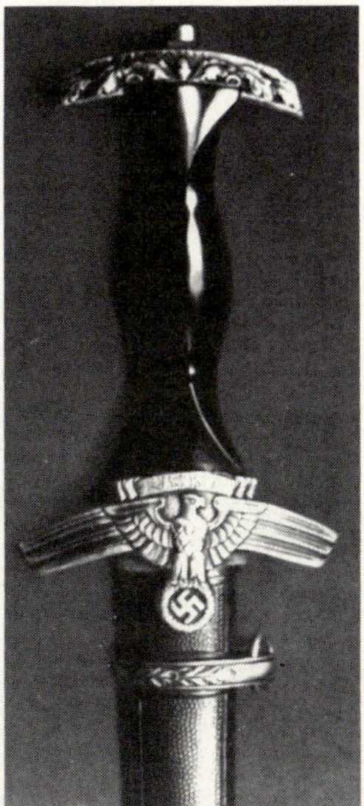

RON FABIAN

THE RAD

The RAD Hewer below, with full stag grip plates, was a product of the lack of standardization during the first months of 1934, after wear of the dagger sidearm was authorized. Inspection criteria was put into effect late in 1934 and such variations soon disappeared. The firm of Karl Julius Krebs produced still another variation towards the end of the war, but prohibition orders on the wear of dagger sidearms after mid-1944 prevented widespread distribution of the design which has become known as the "Fatman". Note the the owner's name is engraved on the lower portion of the ferrule and his initials adorn the reverse of the upper scabbard fitting.

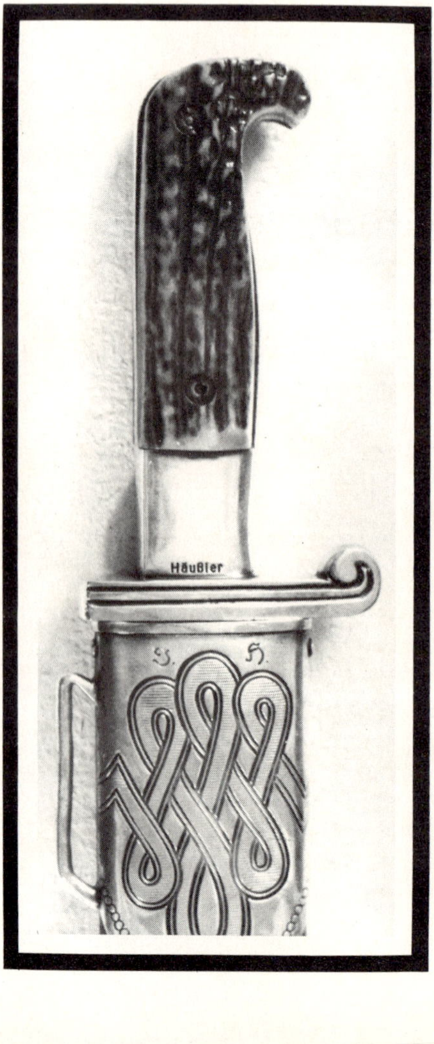

JOHN SCHLEICHER

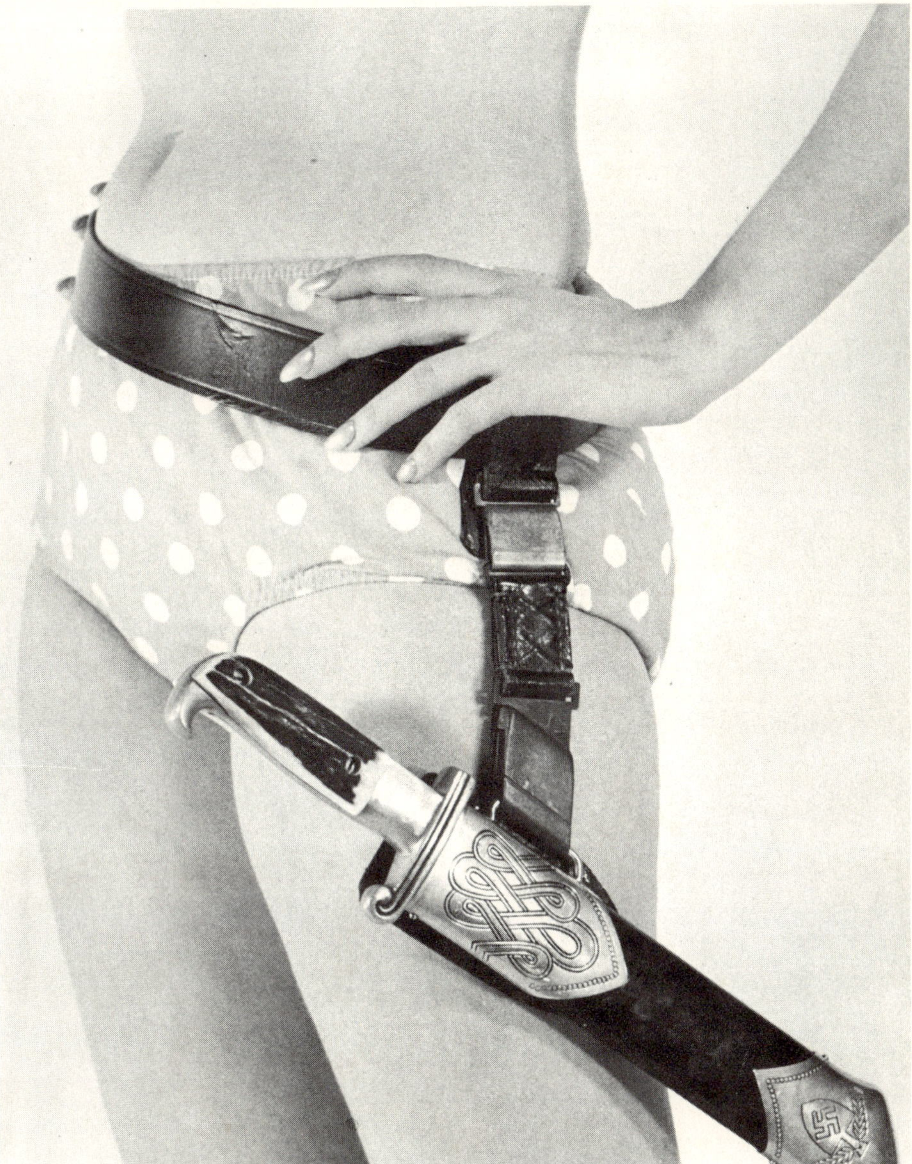

PHOTO: ANDY SOUTHARD

THE LABOR CORPS SUBORDINATE'S HEWER WORN IN THE MANNER ONE WOULD IMAGINE IT WOULD ATTRACT THE MOST ATTENTION. NOTE THE COMPLETE ASSEMBLY OF THE SECOND PATTERN SUSPENSION HANGER.

JOHN SCHLEICHER

RAD LEADERS, WEARING THE 1937 PATTERN RAD HEWER, AWAIT THEIR CALL TO PARTICIPATE IN PARTY DAY ACTIVITIES.

A variation of the standard 1937 pattern RAD Leader's Hewer is illustrated below. The crossguard and ferrule is a solid unit which is heavily silver-plated. The grip plates are translucent plastic simulating mother-of-pearl. All other aspects of the dagger are the same as the basic pattern. This photograph is provided to the reader simply because the specimen exists. Considering the nature of construction, especially of the grip plates, there is some serious doubt as to the originality of the piece. No other specimen of this type is known, and there is no available information pertaining to such a variation.

THE NSFK

Uniform regulations of the NSFK[1] prescribed that the NSFK Flyer's Knife be worn in the following manner:

Full Service Uniform: Authorized; worn with securing strap suspended from outer belt.

Abbreviated Service Uniform: Authorized; worn with securing strap suspended from outer belt.

Walking-Out Dress Uniform: Authorized; worn without securing strap and suspended by means of the suspension strap's bayonet clip hooked under the left tunic pocket flap.

Street Uniform: Authorized; worn without securing strap and suspended by means of the suspension strap's bayonet clip hooked under the left tunic pocket flap (as applies from Sturmführer upward). From Obertruppführer and below, same, but worn suspended from the leather belt without securing strap.

Evening Uniform: Not authorized.

Athletic Uniform: Not authorized.

Ski Uniform: Not authorized.

(ABOVE) NSFK-KORPSFÜHRER CHRISTIANSEN, SHOWN HERE STANDING NEXT TO GAULEITER ADOLF WAGNER, WEARS THE NSFK FLYER'S KNIFE WITH THE SERVICE UNIFORM. NOTE THAT THE KNIFE IS WORN WITHOUT THE SECURING STRAP, WHICH IS CONTRARY TO EXISTING REGULATIONS. (RIGHT) NSFK MEN, PARTICIPATING IN MODEL FLIGHTS, WEAR THEIR KNIFE WITH THE SECURING STRAP.

---

[1] Organizationsbuch der NSDAP, 1937 and 1943 editions.

T. FLURCHICK

THE HITLER YOUTH

THIS HITLER YOUTH KNIFE WAS REMOVED FROM THE WKC FACTORY BY AN AMERICAN VETERAN IN 1945. IT PROBABLY REPRESENTS THE REMAINING ON-HAND STOCK WHEN WEAR OF THE SIDEARM WAS ORDERED SUSPENDED. THE UNUSUAL PATTERN OF THE BLADE WOULD INDICATE THAT FIRMS CONTINUED TO ATTEMPT NEW DESIGN INNOVATIONS TO ATTRACT POTENTIAL SALES.

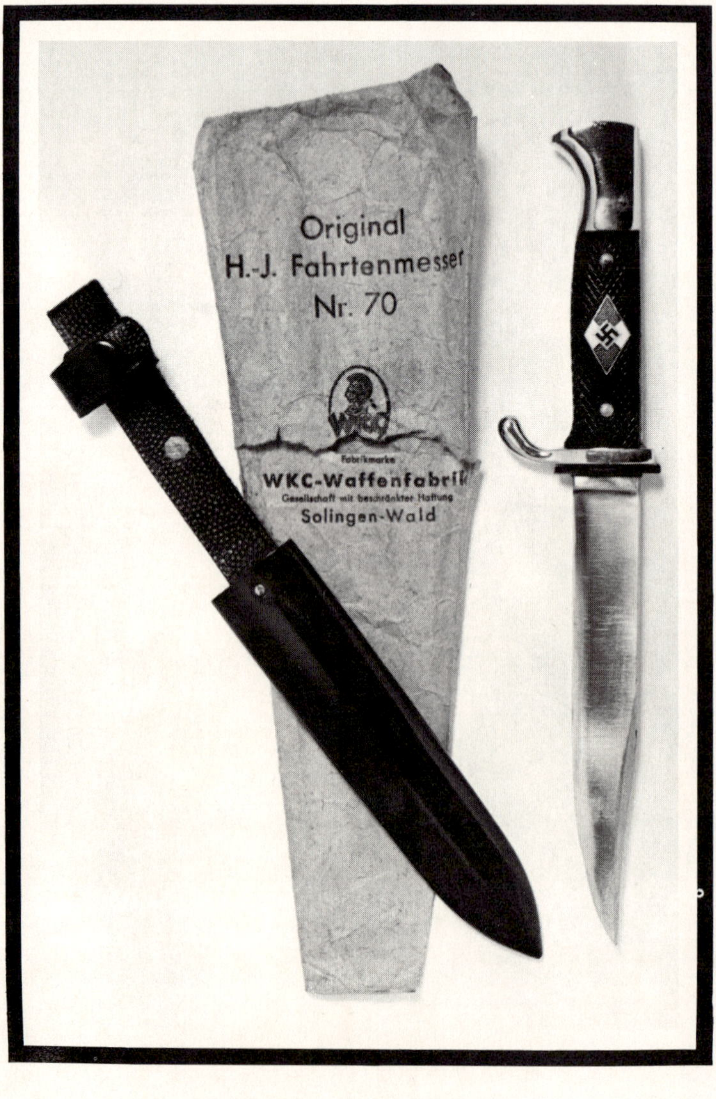

PHOTO: ANDY WALKER

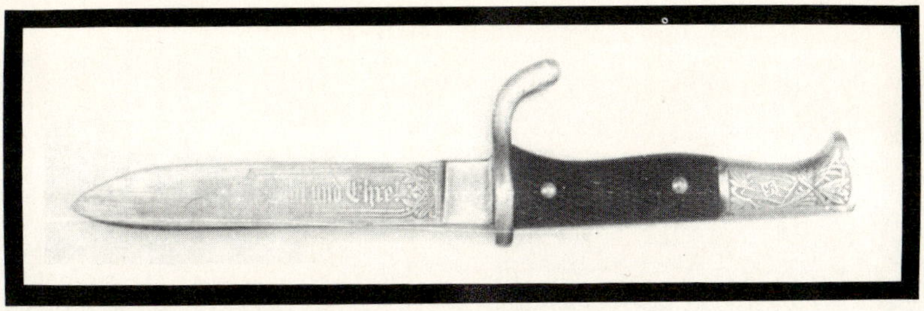

PHOTO: FRED STEPHENS

VARIOUS DESIGN MODIFICATIONS OF THE HITLER YOUTH KNIFE WERE PRODUCED TO INCLUDE ONE WITH A BLADE BEARING THE SS MOTTO, "MEINE EHRE HEISST TREUE". THIS PARTICULAR PATTERN, BASED ON MANUFACTURING TECHNIQUES, APPEARS TO HAVE BEEN AN EXTREMELY EARLY PIECE AND PROBABLY DID NOT SURVIVE AFTER 1936.

A HITLER YOUTH MEMBER, AMONG JAPANESE GIRLS, HAS HIS PHOTO TAKEN LIKE ANY OTHER TOURIST WHILE ATTENDING A CEREMONY IN TOKYO, JAPAN, IN 1939. NOTE THE WHITE LEATHER ACCOUTREMENTS TO INCLUDE THE SUSPENSION FOR THE YOUTH "FAHRTENMESSER".

HITLER YOUTH LEADERS ENDURE A SNOW STORM ON MARCH 16, 1939, DURING A COURTESY VISIT AT OBERSCHLEISSHEIM. THE HITLER YOUTH LEADER'S DAGGER IS WORN SUSPENDED FROM THE BASIC BLACK LEATHER HANGERS, ATTACHED TO A RING UNDER THE LEFT OVERCOAT POCKET FLAP AS PRESCRIBED BY REGULATIONS.

REICHSJUGENDFÜHRER BALDUR VON SCHIRACH AND OTHER NAZI LEADERS GREET HITLER UPON HIS ARRIVAL AT AN INFORMAL RALLY IN MUNICH.

(ABOVE) A HITLER YOUTH CONTINGENT REPRESENTING GERMANY AT A CEREMONY IN TOKYO IN 1939. NOTE THAT SOME YOUTH LEADERS WEAR THE YOUTH LEADER DAGGER WITH WHITE HANGERS AND WHITE LEATHER OR SILVER SCABBARDS, WHILE OTHERS WEAR THE BASIC "FAHRTENMESSER", ALSO WITH WHITE LEATHER SUSPENSION. (RIGHT) LEADER OF THE HITLER YOUTH CONTINGENT ADDRESSING THE AUDIENCE.

PRIVATE COLLECTION, UK

PHOTO: MARC BROWN/KEN SCHINDLER COLLECTION

Illustrated above is a very unusual Hitler Youth sidearm of unknown origin or purpose. The hilt is constructed of a one-piece casting of medium weight alloy, and nickel-plated. The grips are checkered plastic, and are of the same size as the standard "Fahrtenmesser". The blade is of the RLB style, but has the Hitler Youth motto, "Blut und Ehre" etched on the obverse. The manufacturer is Christianswerk, Solingen. The overall length is 13 1/2 inches, while the blade measures 8 3/4 inches. Unfortunately, there was no scabbard present when the piece was discovered. No additional comment is offered, but the photo is presented for the readers' interest.

THE NPEA

The NPEA Service Dagger or "NPEA-Kursteilnehmer Dienstdolch" was first worn by instructor/staff members, but soon was carried as a part of the regulation dress by at least some student members. The specimen below is rather unique since the suspension frog and scabbard are black instead of the usual brown or olive drab. The obverse of the blade bears the standard motto, "Mehr sein als scheinen" (Be more than you appear to be). This is one of the few specimens which bears the manufacturer's trademark (WMW) rather than the distributor's mark of Karl Burgsmüller, Berlin. The only other such blade designation was produced by the Eickhorn firm.

PHOTO: ANDY WALKER

"DAS ILLUSTRIERTE BLATT", FRANKFURT, 1938

AN NPEA INSTRUCTOR OBSERVES THE MARCHING FORMATION OF A GROUP OF STUDENTS AT THE POTSDAM "GROSSE WAISENHAUS" IN 1938. NOTE THAT THE STUDENTS WEAR THE STANDARD HITLER YOUTH "FAHRTENMESSER" WHILE THE INSTRUCTOR WEARS THE NPEA SERVICE DAGGER.

FOLLOWING THE PARADE DEMONSTRATION, THE STUDENTS UNDERGO A RIGOROUS INSPECTION. BASED ON THE DARKNESS OF THE LEATHER, IT APPEARS THAT THE SUSPENSION FROG AND POSSIBLY THE SCABBARD IS BLACK RATHER THAN BROWN OR OLIVE DRAB.

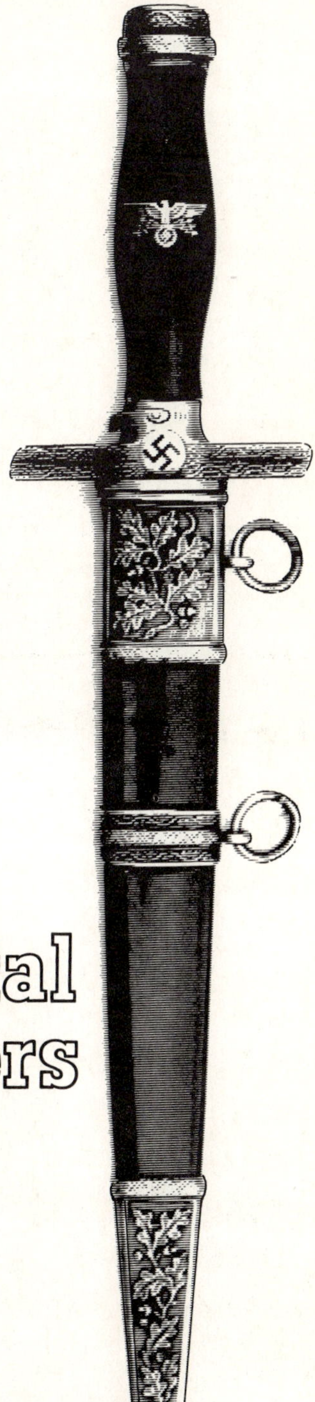

Governmental Daggers

POSTAL PROTECTION FORCE

UNIQUE PATTERN OF THE POSTAL PROTECTION LEADER'S DAGGER WHICH HAS BEEN ORNATELY DETAILED WITH AN OAKLEAF DESIGN. MADE BY WEYERSBERG, THE OBVERSE OF THE DAMASCUS BLADE BEARS THE RAISED INSCRIPTION, "FROM THE POSTSCHUTZ OF THE DISTRICT OF ERFURT". CHAIN NOT ORIGINAL WITH SCABBARD. THE REVERSE OF THE BLADE BEARS THE INSCRIPTION, "TRUTH IS THE MARK OF HONOR".

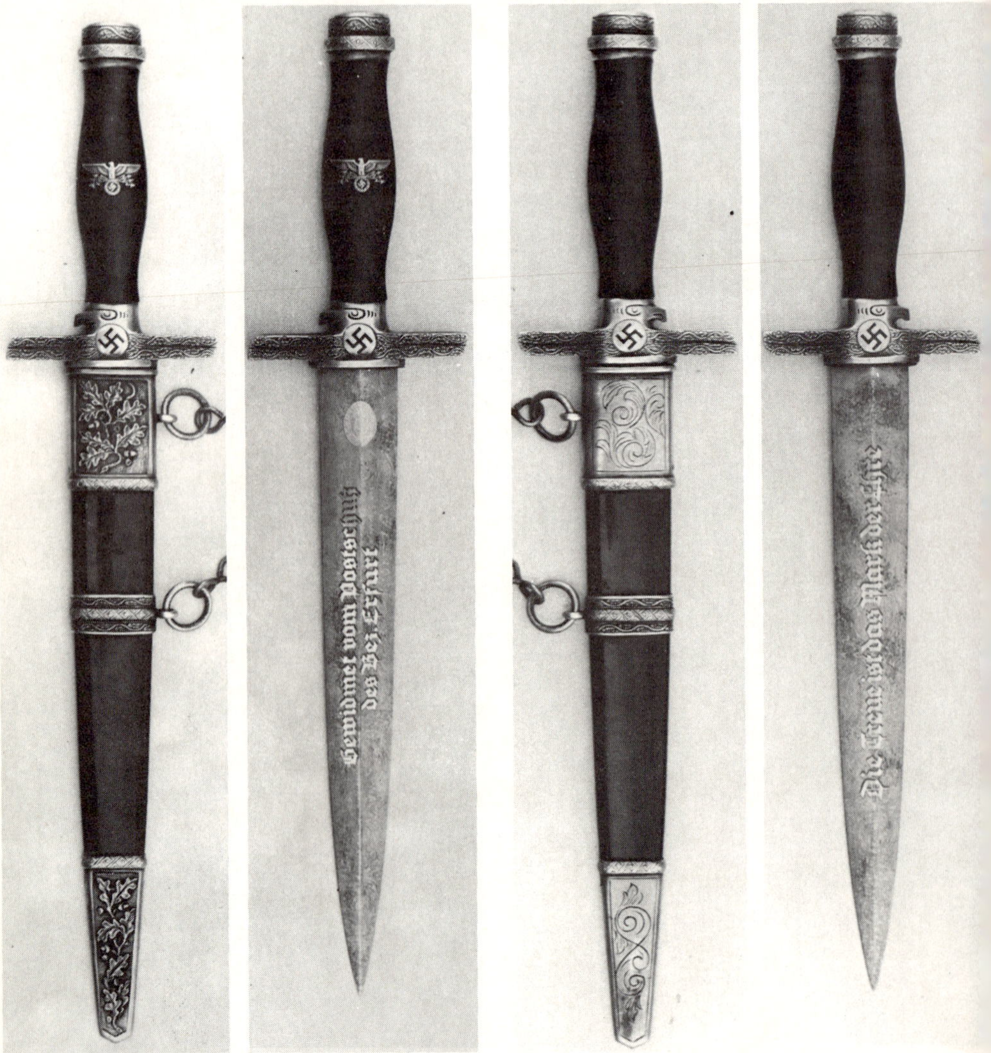

PHOTO: ANDY WALKER/JIM MORRISON COLLECTION

OBVERSE REVERSE

THE RED CROSS

At this time, I would like to correct a statement previously made concerning the manufacture of the Red Cross Subordinate's Hewer. While the Klaas firm did in fact hold a patent pending for a particular aspect of this design, they were not the sole manufacturer of this pattern dagger. As shown here by the initials P.D.L. (P. D. Lüneschloss) at least one other manufacturer produced this pattern as well.

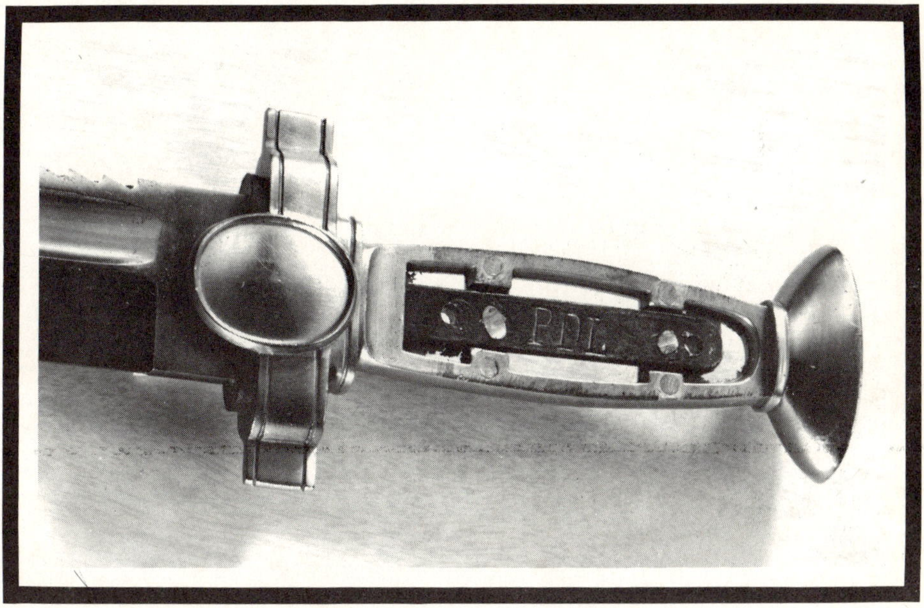

RICHARD SEEBACH

EXAMPLES OF THE DRK SUBORDINATE'S AND LEADER'S HEWERS WITH ORGANIZATIONAL CONTROL MARKINGS ON THE REVERSE OF THE CROSSGUARD.

THE FIRE DEPARTMENT

The Fire Department Official's Dagger below, produced by WKC, has all of the metal fittings finished with a gold plate. This pattern dagger is a carry-over from the Weimar period, and was worn by officials during the Nazi era. This is the only gold pattern Fire Department Official's Dagger known to have existed during this period. Note the fire ladder incorporated into the etched motif on the reverse of the blade.

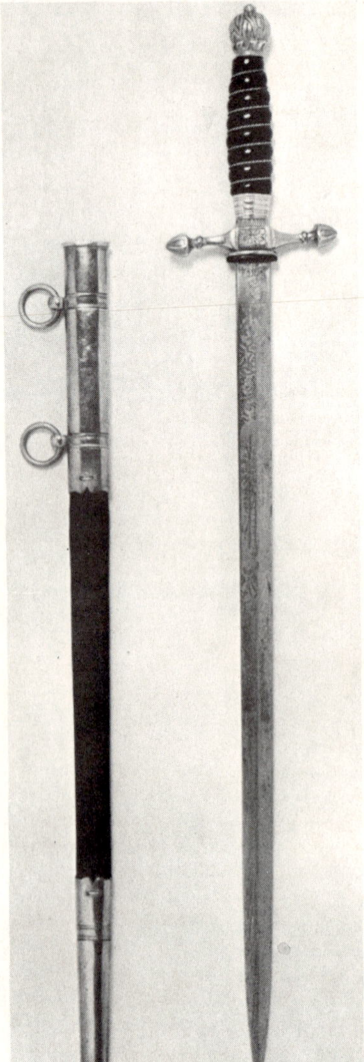

PHOTO: ANDY WALKER

STATE FORESTRY SERVICE

SIDEARM WORN BY LOW RANKING OFFICIALS AND ASSISTANT FORESTERS OF THE STATE FORESTRY SERVICE. THIS F.W. HÖLLER PATTERN HAS THE NATIONAL EMBLEM INCORPORATED INTO THE DESIGN OF THE FERRULE.

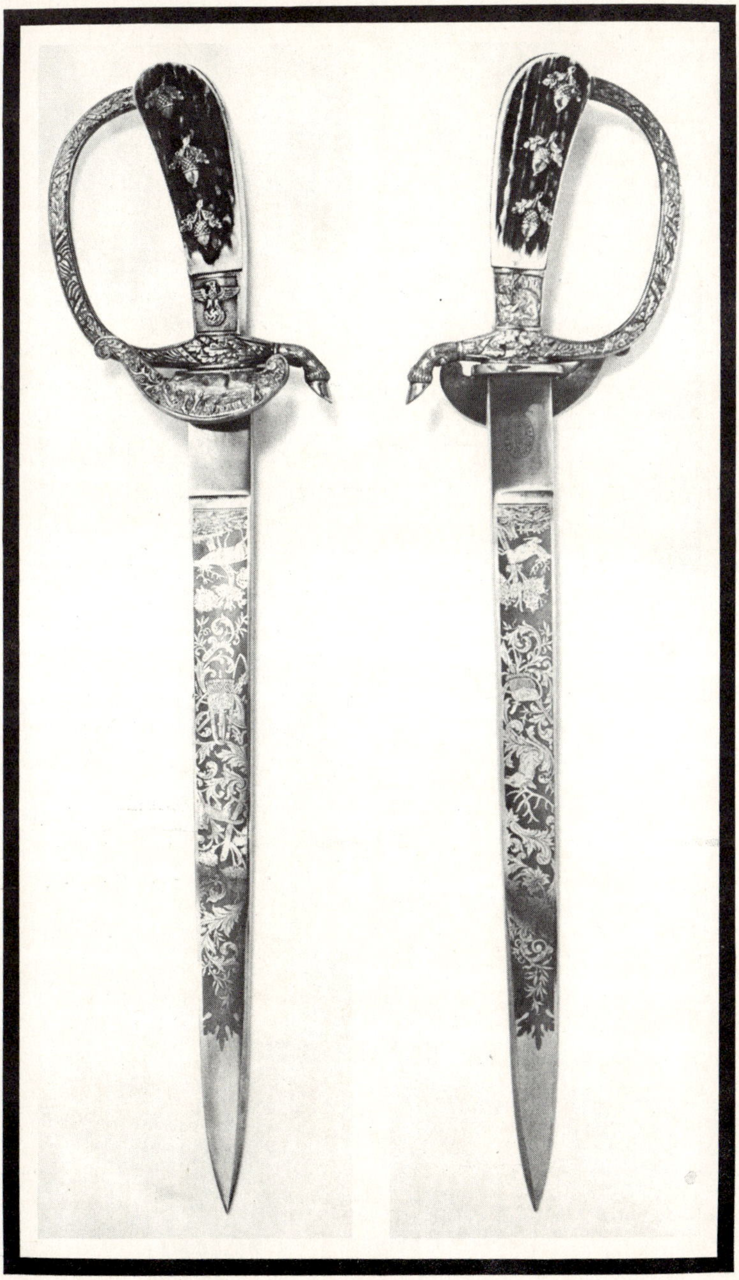

PHOTO: ANDY WALKER

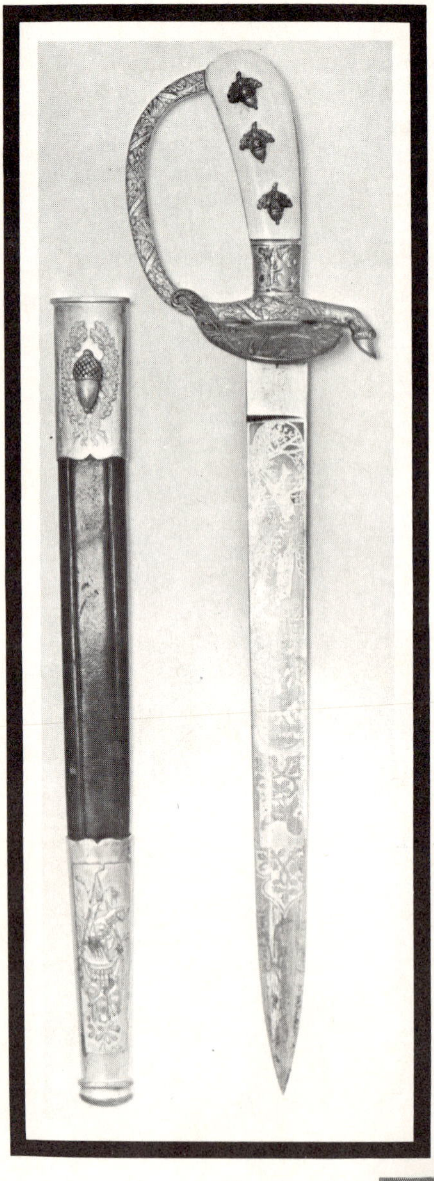

SIDEARM OF THE FORESTERS AND SENIOR OFFICIALS OF THE STATE FORESTRY SERVICE. THIS PATTERN, MANUFACTURED BY THE EICKHORN FIRM, INCORPORATES THE FIRM'S TRADEMARK INTO THE DESIGN OF THE FERRULE.

PHOTO: MARC BROWN/HICKOX COLLECTION

SIDEARM WORN BY FORESTERS AND SENIOR OFFICIALS OF THE STATE FORESTRY SERVICE, AS DENOTED BY THE IVORY GRIP-PLATES INSTEAD OF STAG GRIPS FOR LOWER RANKING MEMBERS. THE GRIP AS WELL AS THE KNOT, AFFIXED TO THE SIDEARM, DESIGNATED THE WEARER'S POSITION.

REVIERFÖRSTER PAUL-RICHART VOGT, AN OFFICER IN THE LUFTWAFFE FORESTRY SERVICE, WEARS THE STATE FORESTRY KNIFE WITH WHITE CELLULOID GRIP-PLATES AND ALUMINUM PORTEPEE. HERR VOGT OPTED FOR THE PURCHASE OF THE DELUXE PATTERN OF HIS KNIFE.

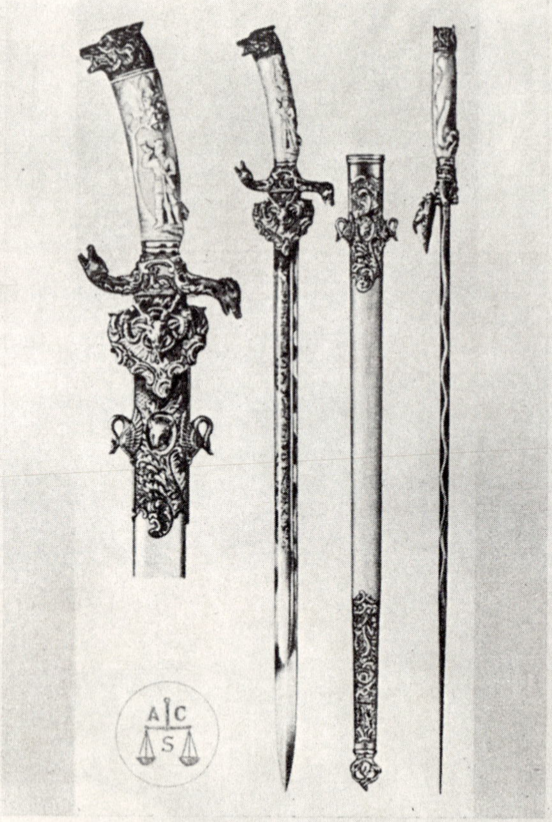

Ein Ehren-Hirschfänger in feinster Ausführung, erst kürzlich noch auf der Internationalen Ausstellung Paris 1937 gezeigt und innerhalb der Ausstellungsgemeinschaft Solinger Stahlwarenfabrikanten mit einer Ehrenurkunde ausgezeichnet!

Er stammt aus der bekannten Fabrik für Alcoso-Qualitätswaffen in Solingen, in der man alle Prunkstücke, aber auch alle anderen Waffen mit der gleichen Liebe und Sorgfalt anfertigt. Auch Ihre Waffe sollte das Zeichen

ALCOSO SOLINGEN

tragen!

"DIE WEHRMACHT", 1938

THIS HONOR SIDEARM OF THE FORESTRY SERVICE WON THE INTERNATIONAL PRIZE IN PARIS, IN 1937. MANUFACTURED BY THE ALCOSO FIRM, SUBSEQUENT PATTERNS MADE USE OF SOME OF THE DESIGN FEATURES WHICH WON THE AWARD.

Boar hunting took on particular thrills when the hunt was consummated by a kill from the hand-held spear. The spearhead below was presented to the Reichsjägermeister, Hermann Göring. Each face of the head's shaft is engraved, the total presentation reading, "Unserm Reichsjägermeister · Hermann Göring · Dauner Werkmannsarbeit · Die Schmiedemeister · Gehendges in Pützborn · Kreis Daun" and ending with a swastika.

"EICKHORN KUNDENDIENST"

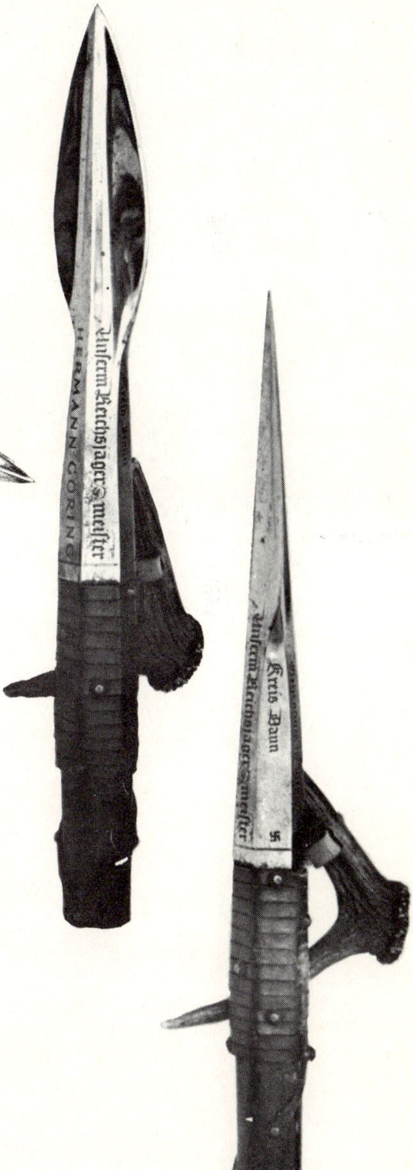

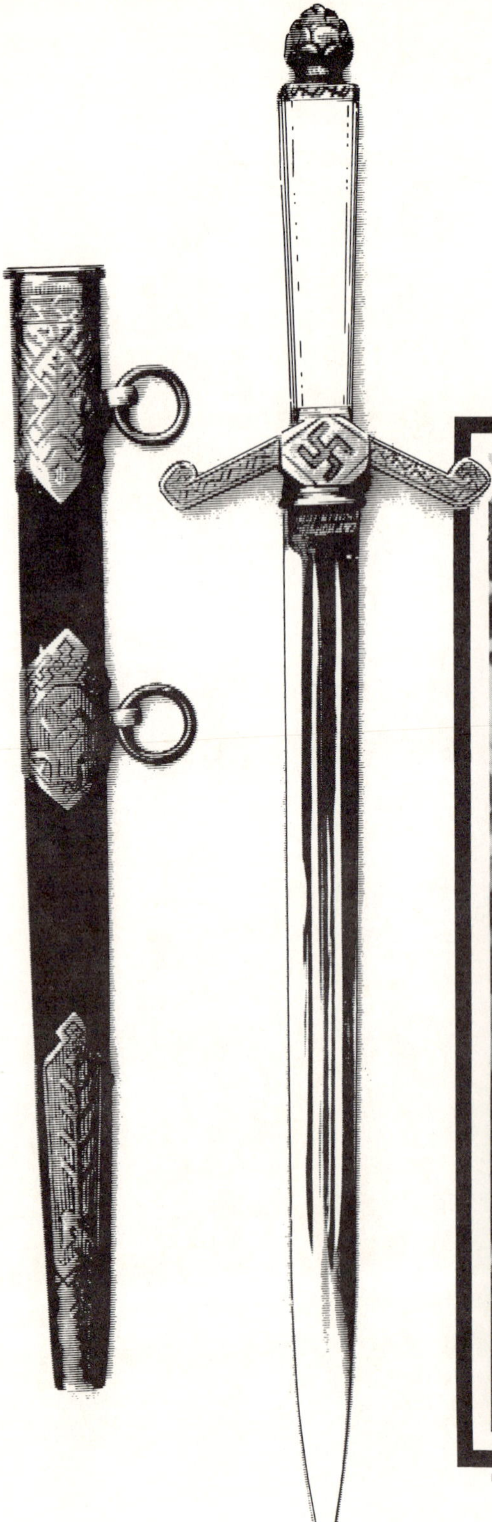

NATIONAL ARCHIVES

Blades of the Axis

CROATIAN COLONEL KREN BEING GREETED IN GERMANY.

It is not the purpose of this chapter to provide a complete and detailed coverage of the daggers and sidearms worn by members of the Axis satellites. It is, however, designed to provide the reader with a representative selection so that an appreciation for the influence of Nazi blades on foreign military and political organs might be gained.

Virtually all European nations which allied themselves with Nazi Germany, either directly or indirectly, provided some form of dagger or sword sidearm for wear by members of their military or political organizations. Spain and most of the eastern European countries manufactured their own specimens, while the countries bordering immediately on Germany largely purchased their sidearms from firms within Germany. Although the designs were unique for the individual countries, the German influence crept into the final design when such countries as Holland, Latvia, Rumania and Croatia purchased their daggers from the Solingen firms.

Some Axis countries could boast a history of blades pre-dating that of Germany, but it was the Germans who used the blade so successfully as a motivating device. Daggers were worn mainly by officers of the military services, but the practice spread to the political organizations as well. Italian Fascists, for example, were wearing dagger sidearms long before the first Nazi dagger was ever designed.

Italy was second only to Germany in the manufacture of blade sidearms. A wide variety of specimens were available for wear by the various Italian organizations. Considering the number involved, and a previous study available, no effort will be made to discuss the basic Italian sidearms as they would involve a major research effort in itself.

With the exception of the daggers produced in Solingen for export, daggers worn by the Axis countries were relatively crude when compared with the Nazi blades. The designs were often progressive and appealing to the eye, but the subsequent manufacture left a great deal to be desired. Eastern European firms simply did not have the expertise boasted by the Solingen firms. Even the Italians, who produced daggers in relative

HITLER IN ROME IN MAY 1938.

quantity, did not have the quality production which would be expected after so many years of manufacturing. One must acknowledge that the Italians did improve in their techniques, and the final results did show vast improvements. Still, compared with the German blades, the Italian's remained second best.

AUSTRIA

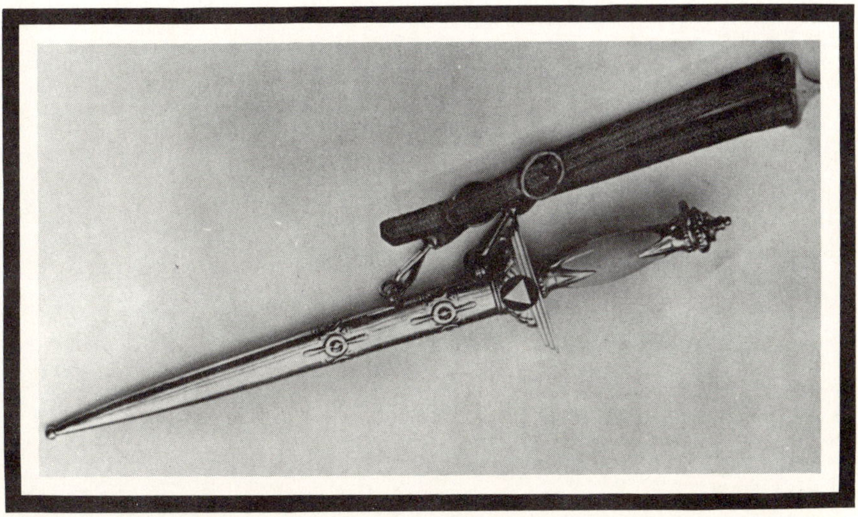

OFFICER'S DAGGER OF THE AUSTRIAN REPUBLICAN AIR FORCE WITH DOUBLE SUSPENSION HANGER.

CROATIA

The Croat Leader's Dagger below is worn by the Croat leader himself, Dr. Ante Pavelić, during his visit to Germany. The overall appearance of the piece lacks the quality or workmanship when compared with the Solingen produced daggers, but is a good example of eastern blade production. The design on the pommel, crossguard and scabbard incorporates a twisted-rope motif. The obverse of the pommel has the initials of the Independent Croatian State (NDH) or "Nezavisna Drzava Hrvatska".

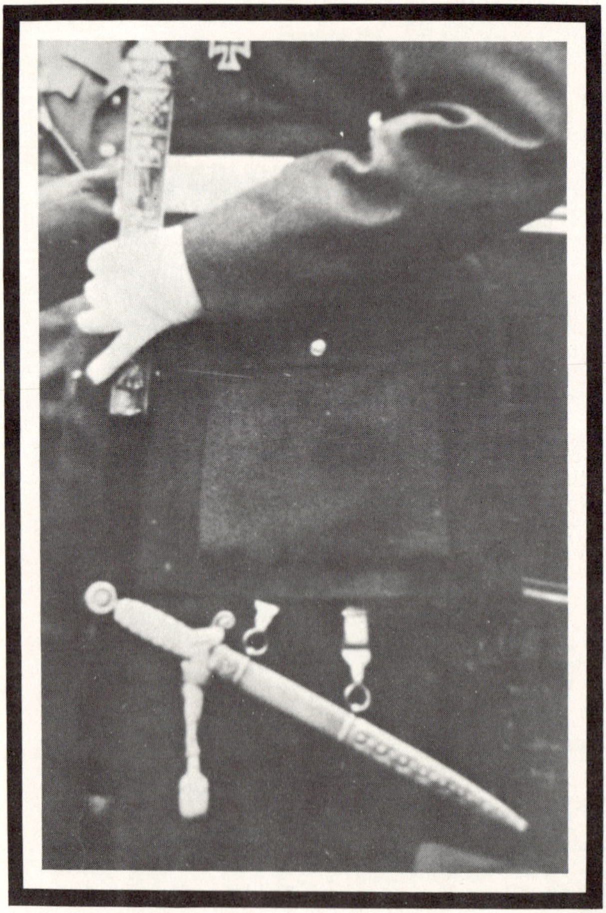
LIBRARY OF CONGRESS

(RIGHT) CROAT AIR FORCE OFFICERS WEAR THE CROAT LEADER'S DAGGER SUSPENDED FROM A DOUBLE SUSPENSION HANGER. RATHER THAN BEING AN OPTIONAL ATTACHMENT, IT APPEARS THAT THE PORTEPEE WAS AN INTREGRAL PART OF THE DAGGER.

FINLAND

Illustrated below is a special presentation dagger awarded to Baron Carl Gustof Emil Mannerheim following the signing of the German-Finnish Pact. The design is probably unique and the dagger one-of-a-kind. Note the use of the Finnish swastika on the crossguard rather than the Nazi swastika, which stands on end.

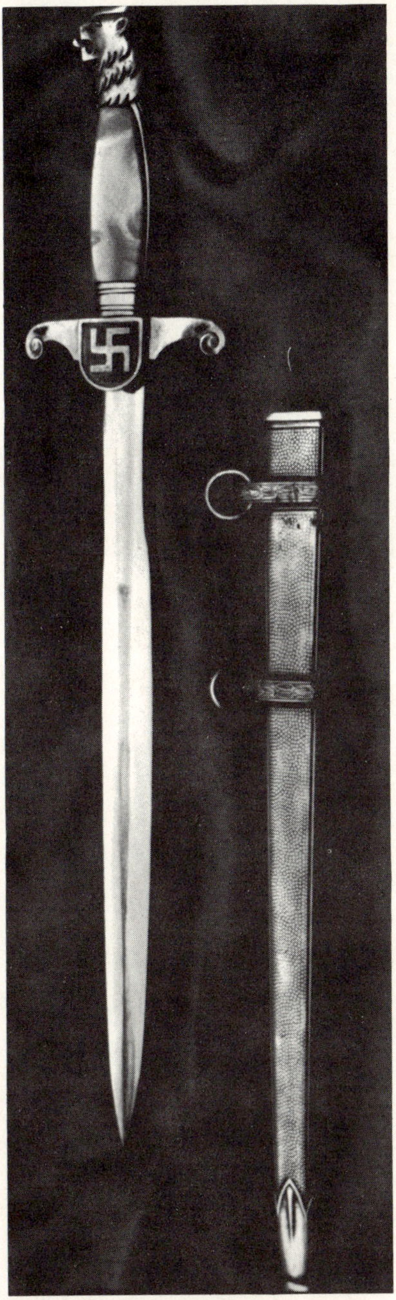

WAGNER-PIAR COLLECTION

HOLLAND

The illustrated officer's dagger of the Dutch Nazi Party, was manufactured by the Eickhorn firm and exported to Holland. Its blade bears a striking resemblance to the RAD Leader, but is without etching or motto of any type. The only blade marking is the Eickhorn trademark on the reverse. A tang nut in the pommel secures the blade to the handle in the same manner as found on the Diplomatic Dagger. The piece measures 36.8cm in overall length. The scabbard is black painted steel with externally mounted nickel fittings. There also exists another design of this same pattern, having a center scabbard fitting and being suspended by a double strap hanger rather than the vertical suspension.

It should be noted that a dagger of identical design, but with crossed hammers on the crossguard, was designated in the 1941 Höller catalog as the Leader's Dagger of the German Miners Union.

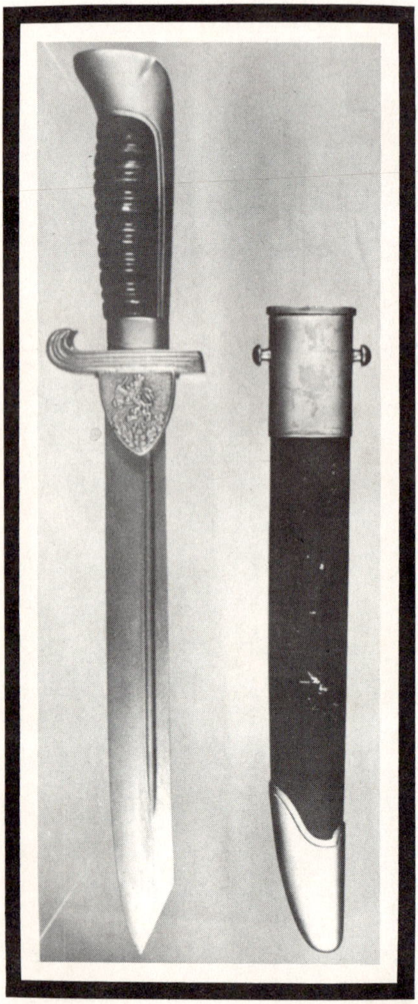

PHIL STOBER

OFFICER'S DAGGER OF THE DUTCH NAZI PARTY.

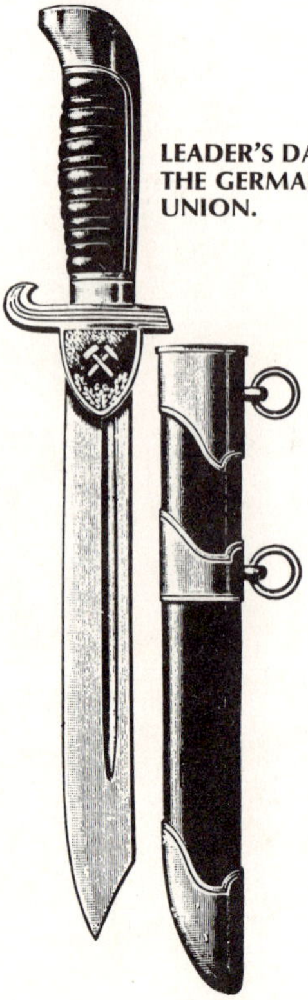

LEADER'S DAGGER OF THE GERMAN MINERS UNION.

On the Dutch Youth Knife, the hilt is heavy nickle-plated steel with a serial number stamped on the reverse of the crossguard. The grip is ivory with an enamel inlay of the Dutch Youth centered in the obverse grip plate. The grip plates are secured by two nickle-plated steel rivets. The blade is smaller than the basic pattern Hitler Youth and its obverse bears the Dutch Youth motto, "Moed, Eer en Trouw", while the reverse is etched with the Eickhorn trademark, the only known manufacturer of this pattern knife. The scabbard is basically identical with the Hitler Youth Knife... suspended by a black leather belt loop affixed to a black enameled steel scabbard.

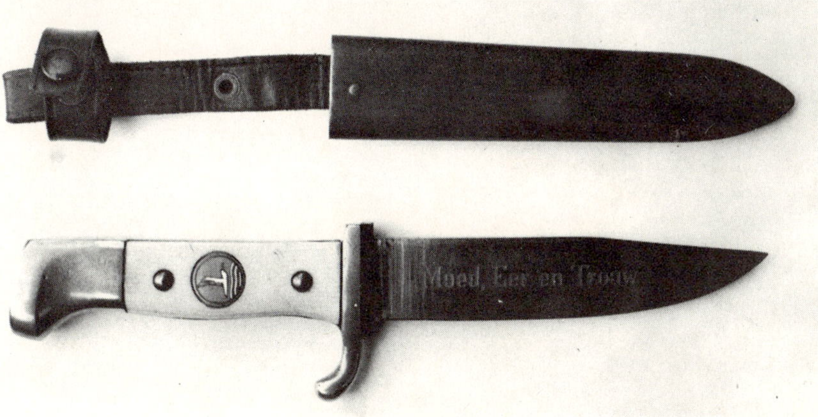

PHOTO: FRED STEPHENS

THE DUTCH YOUTH KNIFE.

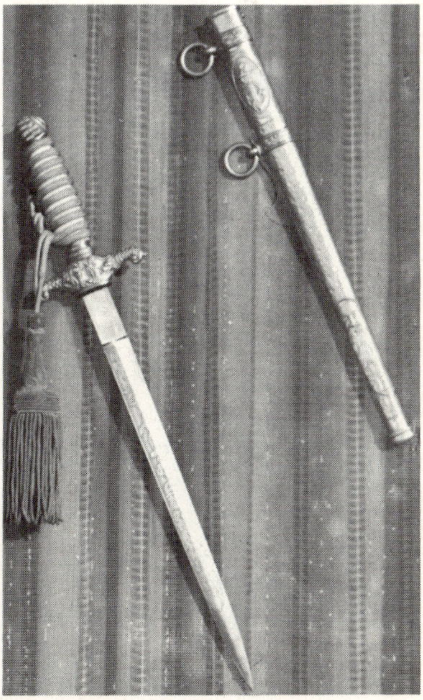

DR. JULIAN MILESTONE

STANDARD PATTERN DUTCH NAVY OFFICER'S DRESS DAGGER. NOTE THE UNUSUAL MANNER IN WHICH THE PORTEPEE IS WRAPPED ABOUT THE IVORY GRIP.

HUNGARY

On the Hungarian Air Force Officer's Dagger, the scabbard is moulded brass centering on the Hungarian coat-of-arms. The brass eagle crossguard is supported on a chrome metal bar. The grip is black plastic. Note the issue number 167 on the ricasso indicating that this dagger was installation property, and did not belong to the wearer.

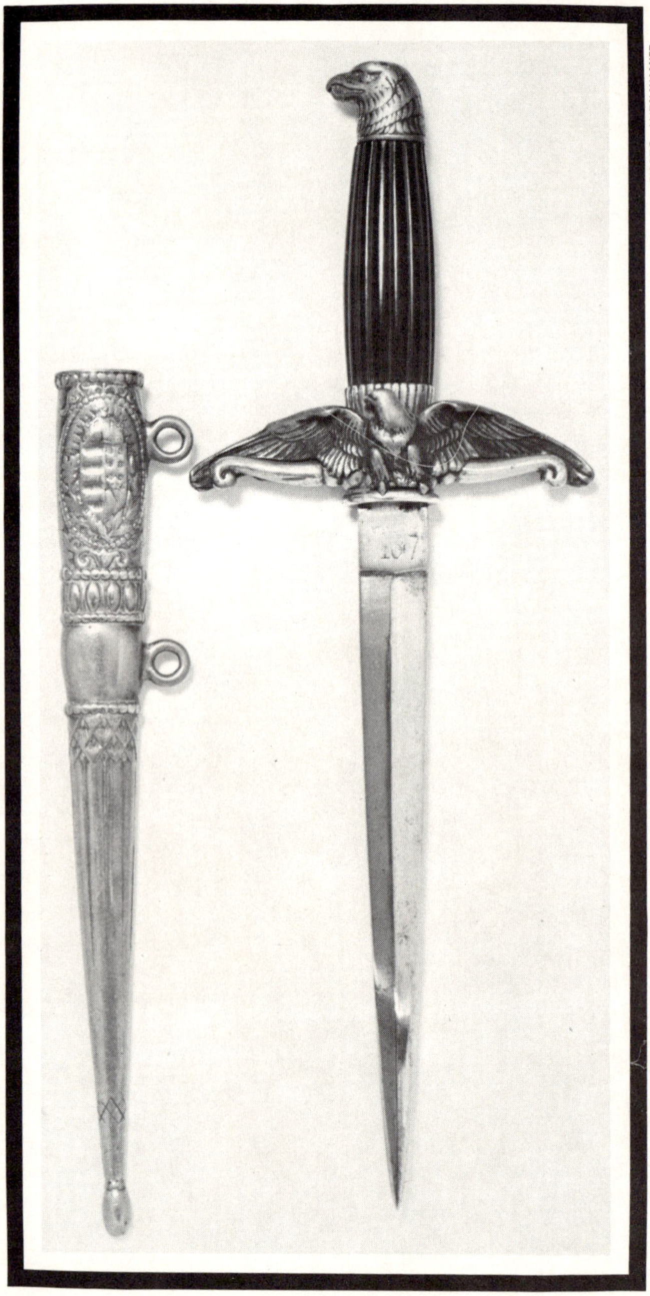

PHOTO: ANDY WALKER

LATVIA

The Army Officer's Dress Dagger of Latvia, illustrated below, was produced by the firm of E & F Hörster in 1938. The pattern is known to have slight variations in design, and at least one other manufacturer... ...Carl Eickhorn. Metal hilt and scabbard fittings are brass, the grip white plastic and the scabbard body leather. The close-up shows an identical pattern dagger, but with a variation pattern blade with etched design...also manufactured by the Hörster firm.

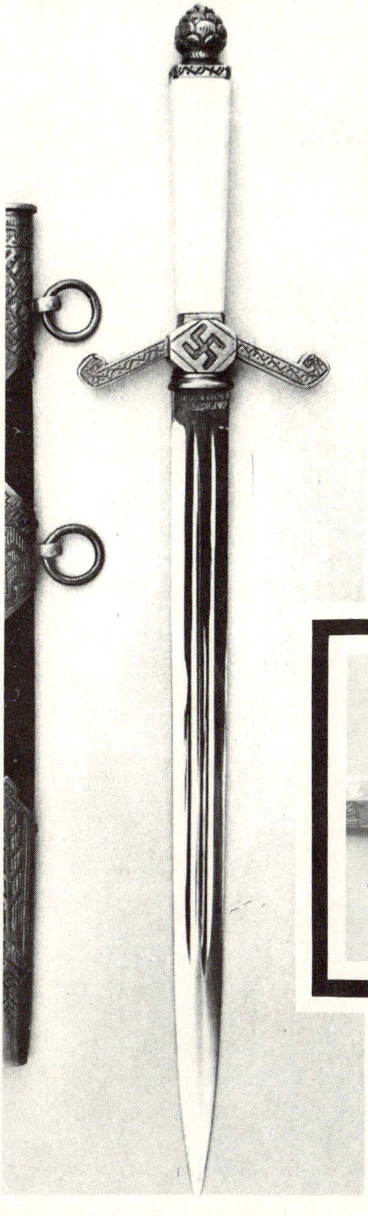

PHOTO: FRED STEPHENS

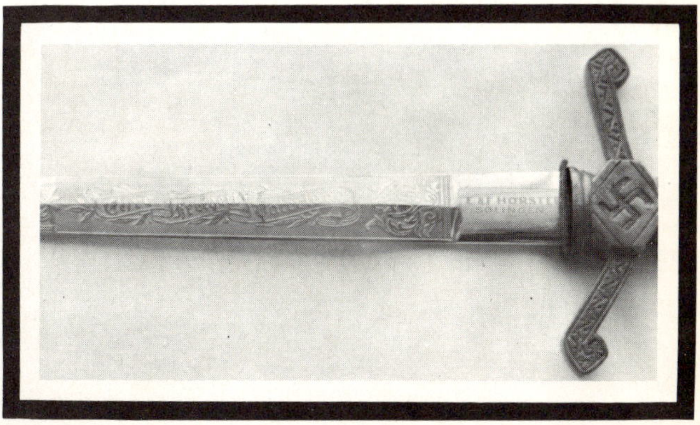

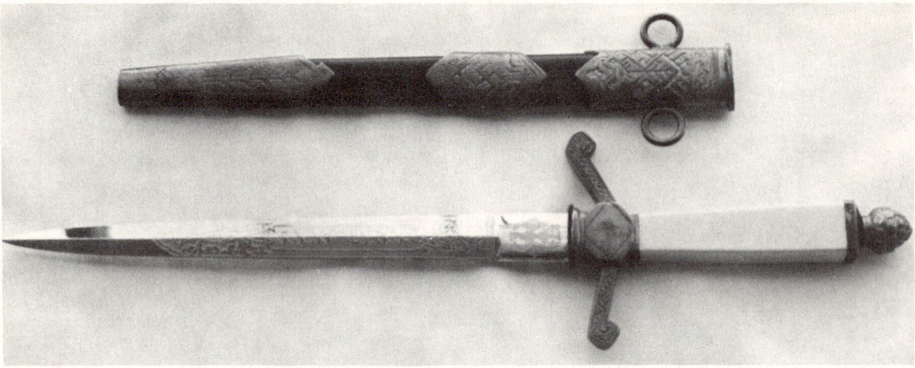

The variation pattern Latvian Officer's Dress Dagger above, with etched blade, shows the etched dates, "5 . 3 . 33 - 21 . 3 . 33". The configuration of the hanger rings would indicate that this dagger was worn in the vertical position. Note the early pattern national emblem on the ricasso.

RUMANIA

The illustrated Rumanian Air Force Dagger was manufactured by the Eickhorn firm, but with the exception of the Navy pattern blade, the design does not bear any resemblance to any Nazi production dagger. The hilt and scabbard fittings are in aluminum and the scabbard body is covered with blue leather. The grip is white celluloid over wood, with brass twist wire wrap. A second pattern of this dagger was produced.. ... identical in every way except that it had a center scabbard suspension band to accommodate a double suspension hanger.

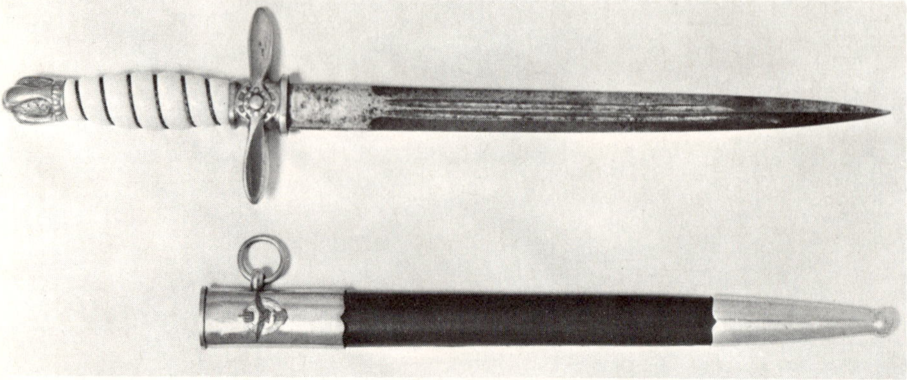

SLOVAKIA

The Slovak leader addresses an assembly in Prague with Nazi officials in attendance. Insufficient information concerning his sidearm is known to warrant comment.

SLOVAK LEADER'S DAGGER

SPAIN

The grip of the Spanish Army Officer's Dress Dagger is reddish-brown wood with brass pommel and crossguard, and the scabbard is simular to the SA pattern. It is believed that a pattern also exists with the chain suspension, thus a scabbard with the center suspension band. All specimens noted have been serial numbered, usually in low two-digit numbers, and the crossguard marked with the "HG". This dagger has erroneously been attributed to the "Eastern Peoples". The accompanying photo clearly shows the dagger being worn by a Spanish Army officer, shown in 1938 talking with members of the Legion Condor.

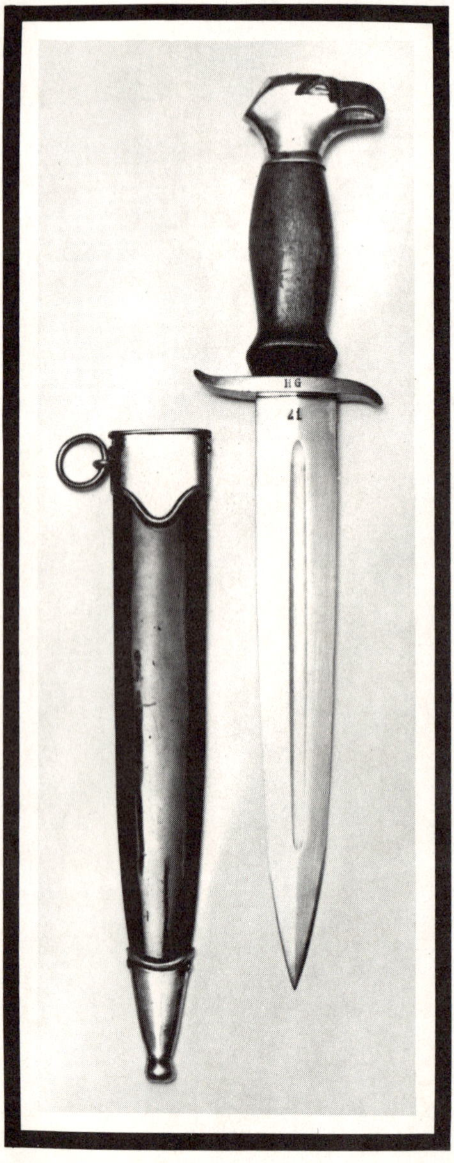

NATIONAL ARCHIVES

THE SPANISH ARMY OFFICER, SHOWN HERE TALKING WITH MEMBERS OF THE LEGION CONDOR, WEARS HIS BASIC PATTERN OFFICER'S DAGGER. INCLUSION OF THIS DAGGER IS TO DISPELL ANY NOTIONS THAT THE DAGGER BELONGED TO ANY NAZI ORGANIZATION. . .ESPECIALLY THE "EASTERN PEOPLES".

Reproductions

Interest in collecting blades has stimulated considerable trade between collectors and dealers, and among other collectors. As the demand grew, collectors derived an appreciation of relative scarcity of the various patterns as did the dealers. From the resultant demand and available supply, values were established and adjusted accordingly. In some cases, value of extremely limited production blades which were highly sought after, achieved values ranging in the thousands of dollars. Values such as these opened the door to temptation to many less-than-honest dealers and some enterprising collectors.

While some altered pieces had long been available, the major influx of reproduction blades began in 1964. The steady stream since that time seems unending. Depending on the intent, various designations have been accorded reproduction blades .... "reproductions", "parts", "new" and "replicas", just to name a few. The intent of this chapter is to address those blades manufactured after May 1945.

True enough, there were stocks of parts in various stages of manufacture and assembly remaining in the hands of the original manufacturers when World War II drew to a close. Large quantities of these parts were resmelted for future use since the economic condition of Germany in 1945 did not permit the luxury of disposing of potential raw materials. Where quantities were small to begin with, reprocessing was not profitable, and these quantities remained intact in the hands of the manufacturers. Some profit-minded manufacturers met the quest of Allied occupation troops for souvenirs by assembling the left-over stocks into finished blades. Often as not, the finished dagger or sword was made of a composite of parts rather than remaining true to the original design. Rather active trade in such assembled pieces took place between 1945 and 1947, until the demand dimished.

Active trade flourished again in 1954 with the import into the United States of the Model 1937 Luftwaffe, SA, SS and Hitler Youth daggers. With the exception of the Hitler Youth knife, the import pieces were a composite of pre-1945 manufactured parts, and post-war parts manufactured to complete the assemply process. In some cases, parts such as scabbards were completely modified in design and material to cut costs and facilitate a quick turnover. Once again there was no total effort to adhere to the original design specifications.

Importation of original parts daggers into the United States in 1964, and later into almost all the European markets, became very active to meet the growing collector demand. At this late date, it became necessary to produce major parts to complete assembly. As the limited supply was exhausted, and demand remained at a high level, entire daggers were newly manufactured. After 1964, manufacturers made a concerted effort to meet the exact specifications of the original design. New production techniques, improved raw materials, and diminished craftsmanship, however, left their tell-tale signs. Not knowing what to look for, or even suspecting that the daggers were anything less than original, collectors continued to trade in these daggers. It was not until the increasing volume of heretofore rare daggers began to be made available, and the apparant newness, that the collector began to suspect that there might be something wrong with his new purchase. It was the aftermath of this discovery that a great number of "experts" in the field of reproductions began to emerge.

Some observations put forth to the collector were correct. However, many times the observations were incorrect, misleading, or only partially true. The initial expertise was largely derived from observation or comparison, but all to often, the comparisons were made on a very limited basis. No account was taken for the variation that existed within the patterns between manufacturers. The misinformation caused, and continues to cause, collectors to bypass desireable original daggers, and originals to be released because they were identified as "reproductions".

The results of this chapter were derived after years of investigation, considerable information being made available by both U. S. and German agencies, and extensive interviews conducted with Solingen manufacturers and former artisans. New and exotic blades continue to be produced, however, and information regarding these is slow to catch up.

The most difficult piece to identify is the original that has been newly altered with a dedication or presentation. The possibility in this area is endless, and only standard patterns will therefore be addressed.

Since the collector has become more knowledgeable, more deceptive techniques have been introduced to continue the trade in reproductions. Putting a reproduction blade with an original hilt assembly, using chemicals to give the appearance of age, and removal of identifying marks are but a few of the deceptive devices used by unscrupulous dealers and collectors.

Since reproductions became a very lucrative commodity in 1964, they have been produced to the present. Production over this time frame has generated variations even within the reproduction patterns, and hard-and-fast rules no longer apply in absolute identification. However, where standardization in production provides identifying features, or an error was worked into production, a rule is established that will allow the collector to identify a reproduction.

AIDS TO IDENTIFICATION

MODEL 1933 SA SERVICE DAGGER

Those imported during the period 1954 were assembled from original and postwar production parts. The unfinished blades were hollow ground, and etched with the standard SA motto. On the reverse of the blade was etched a cross with double letters "B & A", and the name Solingen. Initial import pieces were stamped with "Germany" to meet U. S. import regulations, but later blades bore no such marking. The crossguards and insignia inlays were original, but the grip was constructed of dark reddish-brown plastic. Blades imported during this period either had no scabbard or were fitted with a leather scabbard. The SA Service Daggers produced in 1964 were a marked improvement, but still bore distinct errors. The newly produced blade was hollow ground with a sharp ridge running down the center. Early blades had the RZM mark on the obverse, but this was later corrected, and placed on the reverse of the blade. The insignia inlay was original, taken from a stock of some 5, 000 original pieces of insignia. However, the wooden grip was newly produced, and upon placement of the insignia, the eagle was exactly centered in the grip rather than being offset towards the lower crossguard. This too was later corrected. A poor fit could be noted between the blade and the curvature of the lower crossguard. In all cases, the eagle inlay was constructed of aluminum. Scabbards were taken from a large quantity of originals, and only aid in identification through their new condition and poor fit with the blade.

MODEL 1933 SA ROHM HONOR DAGGER

The same basic comments apply here as those attributed to the SA Service Dagger. A few notable exceptions were the use of the aluminum eagle rather than the required silver-plated eagle, and the use of a facsimile of the Eickhorn trademark in conjunction with the RZM mark. Of course, the RZM mark could never be found on original Röhm Honor Daggers since Röhm was purged before the RZM came into existence. The initial production run of these daggers numbered forty. The use of a reasonable facsimile template, and the correction of the most outstanding errors made these pieces ever more difficult to detect... this was especially true when the spurious dedication was applied to an original

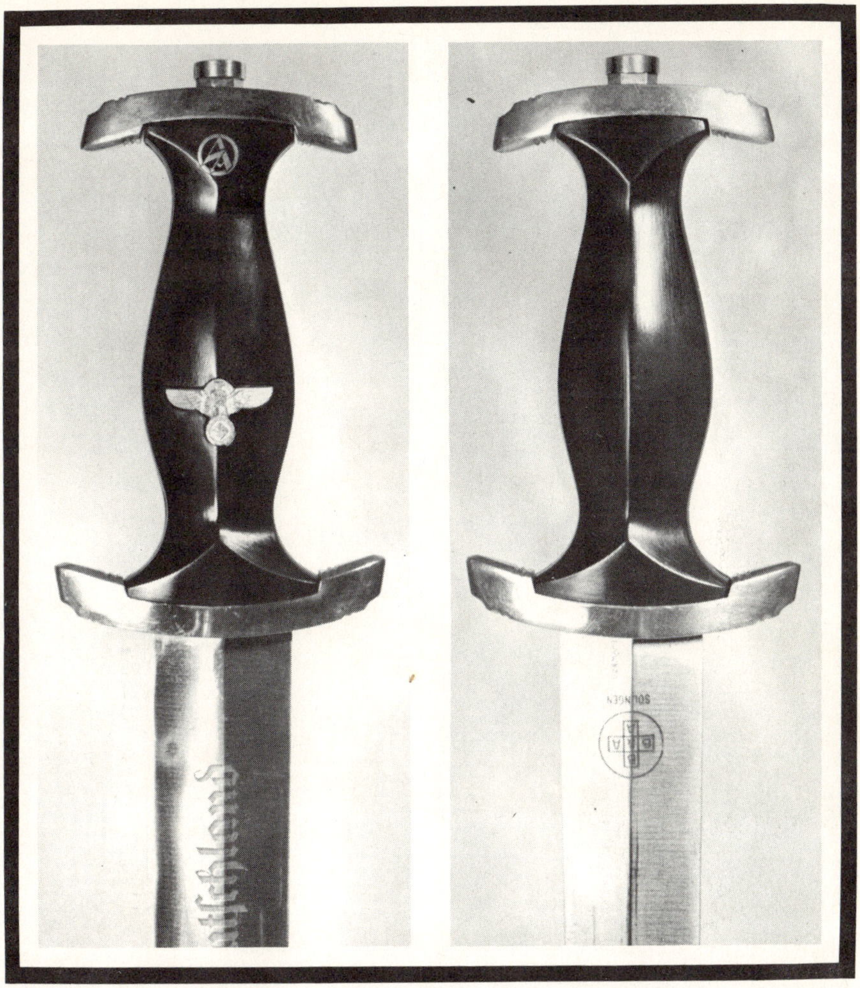

PHOTO: ANDY WALKER

THIS SA SERVICE DAGGER WAS ASSEMBLED FROM ORIGINAL PARTS WITH OTHER NEEDED PARTS MANUFACTURED AFTER THE WAR TO COMPLETE THE ASSEMBLY. SUCH SPECIMENS WERE IMPORTED INTO THE UNITED STATES IN 1954 (NOTE THE IMPORT MARK "GERMANY" JUST ABOVE THE TRADEMARK). ONLY THE INSIGNE INLAYS AND THE CROSSGUARDS ARE ORIGINAL. THE GRIP IS DARK REDDISH-BROWN PLASTIC.

PHOTO: MARC BROWN/NICKOLS COLLECTION

A SMALL QUANTITY OF THESE REPRODUCTIONS WERE MADE AVAILABLE IN APPROXIMATELY 1964. NOTE THAT THE DOUBLE TRADEMARK HAD TO BE POSITIONED VERY NEAR THE CROSSGUARD TO ACCOMMODATE THE PRESENTATION. NOTE ALSO THE FIT OF THE BLADE TO THE CROSSGUARD.

blade. Plated zinc crossguards were normally used rather than the required nickel-plated, early pattern, and did not have the SA group abbreviation.

HIGH QUALITY SA SERVICE DAGGER

This piece was usually achieved through the modification of an original dagger. Various designs were engraved or machine tooled on the upper and lower scabbard fittings, and the upper and lower crossguards. The metal fittings were then often gold-plated. Spurious dedications were engraved or photo-etched on the reverse of original blades as well. Some of the highest quality work produced, originated in Belgium by an expert engraver, and can only be detected under high magnification. A late modification of these pieces was the addition of honor-type crossguards cast with oak leaves and silver-plated. These crossguards were normally silver-plate over brass or some other base metal. Standard pattern Service Daggers with honor-type crossguards are usually assembled to increase the overall value, and must be considered other than original.

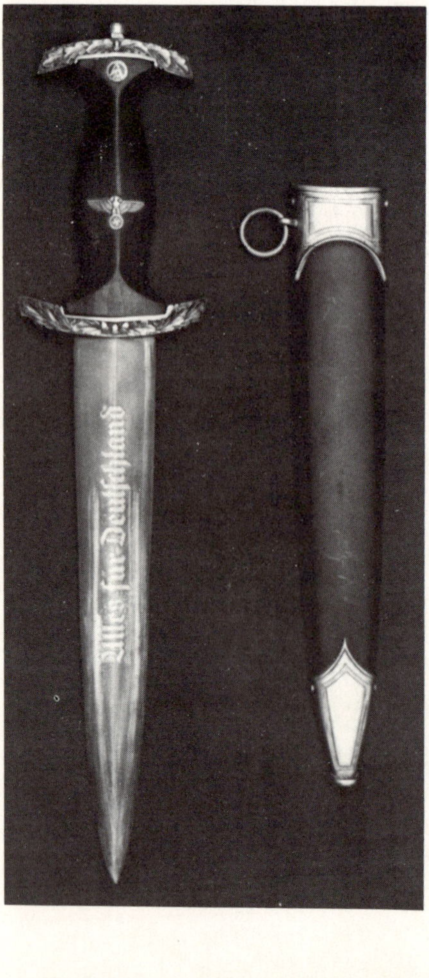

ALBERT GOOSENS

INSTEAD OF BEING CAST, THE CROSSGUARDS AND SCABBARD MOUNTS HAVE BEEN EXPERTLY HAND-ENGRAVED TO GIVE THE APPEARANCE OF THE HONOR-PATTERN FITTINGS. MODIFICATIONS SUCH AS THESE TO A BASIC POLITICAL DAGGER WILL INCREASE THE VALUE FROM APPROXIMATELY $40 TO $1,000.

SA HONOR DAGGER

This particular design has undergone a transition from the very crude to the near perfect duplication. The first patterns to appear were simply modifications of original SA Service Daggers, in that the upper and lower crossguards and scabbard fittings were engraved with an oak leaf motif, and oak leaves were added before and after the motto. It was common practice to gold-plate the scabbard and hilt fittings. A damascus blade of identical design of the original, but lacking the weight and quality, was introduced in 1964. The blade was first manufactured in Germany, in 1963, at the individual unit cost of $20, and was produced by the original artisan, Paul Müller. His increasing age accounts for the tremendous deviation from his original damascus work. The blade was considerably thinner, and of less weight, than the original. The raised motto and oak leaf design on the obverse were lightly gold-plated and were of identical design of the original, but without the depth of the original. A distinct line or ridge ran down the center of the blade instead of being rounded. The blade also bore the marking, "Echt Damast P. Müller", in raised letters, and gold-plated. This blade was first fitted to a standard SA Service Dagger. Early modifications also included a leather covering on the scabbard body. The transition of this piece continued with the addition of honor-type crossguards... a cast oak leaf pattern, which was silver-plated over brass. Later castings were made of solid German silver, but still lacked the quality of detail of the originals. With the training of a new damascus artisan, the quality of the blades vastly improved to the point where only their lack of weight and age can be the distinguishing features from the original. With age, even these blades will become more difficult to detect. Exact reproductions of the scabbard honor fittings were cast from brass and silver-plated. Current modifications include scabbard fittings molded from solid German silver with varying oak leaf designs. Final efforts have resulted in near exact duplications of the originals. The main distinguishing feature is the continued use of varying styles of Müller markings instead of the early or late pattern Eickhorn trademarks. However, this feature is subject to change.

CHAINED SA HONOR DAGGER

The same comments apply as discussed with the latter reproduction of the SA Honor Dagger. Two different designs of the chain suspension hanger were cast in silver... one having the exact design of the original, and the second having a design which was a complete departure from the standard pattern. The most obvious deviation was in the assembly of the individual links, which were assembled in layers rather than produced from a single casting. The variant design consists of a series of insignia...

...usually composed of the swastika, tyrune and SA rune. Facsimile markings were affixed to the reverse of the belt fastener plate and one more on the suspension links.

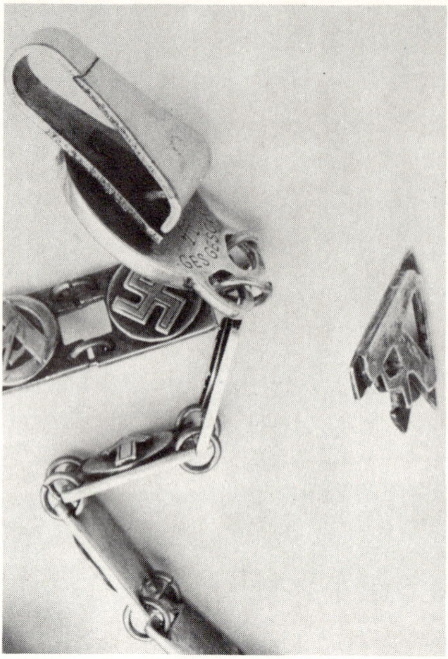

PHOTO: ANDY WALKER/PHIL CAPEWELL

AT FIRST GLANCE, THIS DEVIATION SET OF SUSPENSION CHAINS FOR THE REPRODUCTION CHAINED SA HONOR DAGGER APPEARS TO BE RATHER GOOD. GREAT DETAIL HAS GONE INTO THEIR MANUFACTURE, WHICH IS BELIEVED TO HAVE BEEN ACCOMPLISHED IN ENGLAND, TO THE POINT OF USING THE ASSMANN TRADEMARK. ANY SUCH PATTERN BEARING THIS ASSEMBLY OF FITTING INSIGNIA SHOULD BE CONSIDERED SPURIOUS.

MODEL 1933 SS SERVICE DAGGER

Early patterns of this dagger first appeared in 1964, and were easily identified due to the numerous errors which could be observed. The hollow-ground blade had a sharp line or ridge down the center. The motto was lightly etched, and often off-set towards the blade tip. Early specimens bore the trademark 120/34 with the SS sigrunen above it and the RZM mark below. This was later changed, however, to bear a facsimile of the Eickhorn trademark and the M7 designation with varied manufacturer's code numbers. Initially, the original aluminum eagle was centered in the black dyed, wood handle, but this was later corrected by off-setting the eagle towards the lower crossguard. The metal scabbard was originally taken from a stock of original SA scabbards and painted black. Upon disassembly, it can be noted that the blade tang in many cases has been filed to fit the lower crossguard aperture.

MODEL 1934 SS HONOR DAGGER

Early examples of this dagger were produced from the standard Model 1933 SS Service Dagger reproduction, but with a reasonably good facsimile of the Himmler dedication added to the reverse of the blade. The most obvious error is the use of the dedication on a blade bearing the RZM trademark. This error was later modified by using a genuine SS Service Dagger with facsimile dedication added. Alloy-type crossguards were initially used, whereas the only type of crossguard which could be legitimately found on this dagger was the solid nickle type. Aluminum eagles inset in the black-dyed wood grips were used instead of the correct silver-plated early pattern. Black painted scabbards were used instead of the correct blued metal base scabbard. Original versions of the M34 SS Honor Dagger could be found in a black enamel scabbard providing that a replacement scabbard was purchased, and the change accomplished by the original owner. This was especially true with the advent of the Model 1936 SS Service Dagger. It would have been correct for the original recipient to replace only the M33 scabbard with the newer M36 scabbard, since in many cases, the purchase of a completely new dagger was prohibitive due to cost.

MODEL 1936 SS SERVICE DAGGER

The base dagger and scabbard was nothing more than a standard M33 SS Service Dagger, but with the scabbard modified with a reproduction suspension chain hanger and center suspension scabbard band. The suspension links lacked the detail and were usually unplated. It was not unusual for more than a single link to be marked with the facsimile SS inspection mark or some other characteristic marking. The first chain specimens were non-magnetic, but this was later corrected. Later specimens bore improved blades, with the motto being etched deeper, and the trademark modified to resemble the 1935-1941 Eickhorn trademark. The most noticeable deviation in the trademark was the lack of the sword being held by the squirrel symbol.

SS HONOR DAGGER

The same comments apply for this piece as have been attributed to the SA Honor Dagger. The exception, of course, is in the grip and the blade. The grip inlays were taken from original stock, but the blade was newly fabricated...usually of Turkish damascus. In some cases, the SS sigrunen insigne was affixed to the face of the pommel button.

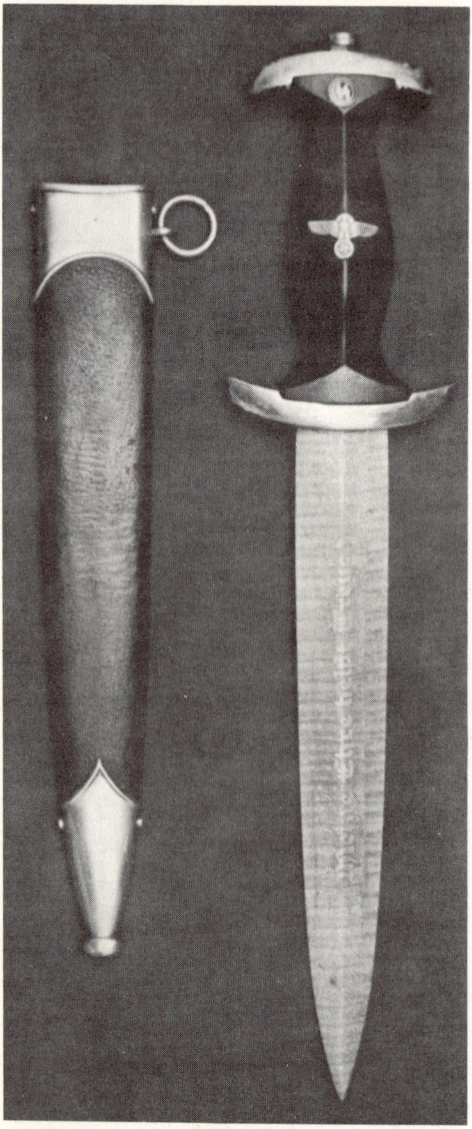

ALBERT GOOSENS

THIS SS HONOR DAGGER WITH DAMASCUS BLADE FIRST MADE ITS APPEARANCE IN 1964. THE MAIN FEATURES WHICH CAN EASILY IDENTIFY THE PIECE ARE THE SHARP RIDGE DOWN THE CENTER OF THE BLADE, AND THE CENTRAL POSITIONING OF THE EAGLE INLAY CAUSING IT TO APPEAR HIGH IN THE GRIP.

MARINE SA SERVICE DAGGER

This piece is a complete fabrication in the pattern used, as the Marine SA did not employ the use of gold-plated dagger fittings. The specimen used was normally an original SA Service Dagger, but with the grip dyed blue, the addition of a blue leather covering on the scabbard, and all metal on the hilt and scabbard gold-plated.

NSKK SERVICE DAGGER

A standard SA Service Dagger reproduction was used, but the scabbard body was painted black.

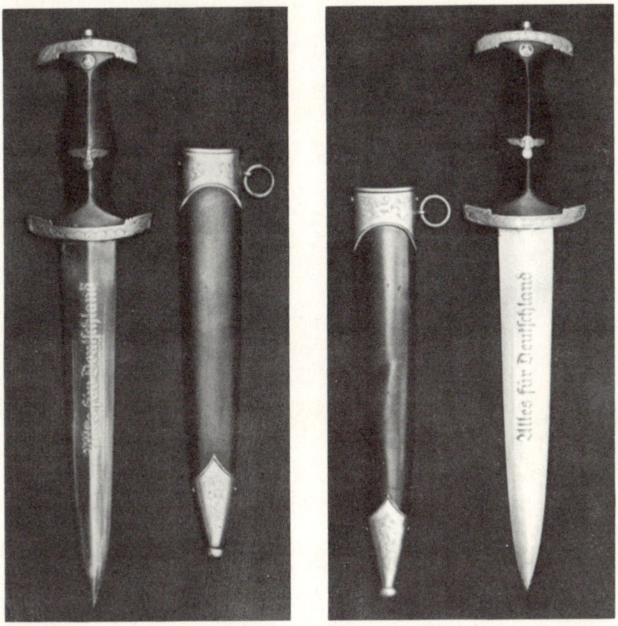

ALBERT GOOSENS

TWO SPECIMENS OF ORIGINAL NSKK SERVICE DAGGERS BEING ALTERED BY ENGRAVING TO CREATE A HIGH QUALITY SERVICE DAGGER.

MODEL 1936 NSKK SERVICE DAGGER

This was an excellent reproduction. especially after an original dagger was transferred to a reproduction scabbard. The primary mistake made was the stamping of "Musterschutz NSKK-Korpsführung" on more than one link of the chain suspension hanger. The Models 1933 and 1936 NSKK Service Dagger have been gold-plated in some cases to represent the legitimate NSKK Marine dagger. The original specimen was copper-plated, however, and not gold-plated as used with the fake specimens.

MODEL 1934 RAD HEWER

The specimens reproduced in limited quantity was that which has become known as the "RAD Fat Man". Mr. Carl Krebs of the Krebs firm, Solingen, readily admitted that this hewer was reproduced at the request of a prominent dagger authority in 1964. However, due to late war production of the original pattern, it is almost impossible to distinguish the post-war reproduction from the war-time counterpart. Identification is further complicated since the Krebs firm used original scabbards and hilt fittings to fabricate the post-war hewers.

MODEL 1937 RAD LEADER'S DAGGER

This pattern was produced by the Krebs firm as well, but a telltale error was committed in production which allows for easy identification. Early specimens were produced with very thin pressed steel blades and were without the blade fuller. While some patterns were produced without any motto, this was later corrected but with a thinly photo-etched motto lacking virtually any depth. The grips were constructed of white plastic rather than the correct celluloid. Original scabbards were used to complete the fabrication.

HITLER YOUTH KNIFE

It would seem that a knife so readily available and of little relative value would not be economically feasible to reproduce. Yet, the knife is still produced in quantity today. However, the intent of those knives produced today is simply for use as a commercial scout or hunting knife with the design taken from the original molds. However, the diamond commercial insigne has been replaced with an original Hitler Youth insigne, and in some cases, the Youth motto etched on the blade. Early specimens had the motto etched on the reverse rather than the obverse of the blade. Thus, the motto and the trademark ended up on the same side of the blade. This was quickly corrected. Emory cloth was taken to the blade to dull the new finish, and to give the knive an aged appearance. Any Youth Knife having the Youth diamond and bearing the "Hubertus" trademark can be considered as a postwar production specimen. A large quality of original Youth Knives were left remaining in stock when the war ended. These were exported to the United States, and were usually stamped with the export mark "Germany" to conform with U.S. custom regulations. While the stamping may detract from the blade, the Knife should not be considered a reproduction. Postwar copies can be identified by means of the hollow-ground blade, and the flattened grip retaining pins on the reverse side of the grip.

HITLER YOUTH LEADER'S DAGGER

At first glance, this reproduction is an excellent specimen, but upon closer inspection, there are a number of errors which are readily detected. Early specimens used fiberglass to form the scabbard body rather than metal. This can be seen as a bluish-grey substance by looking into the scabbard throat. A quick manner of detection is to apply a magnet to the leather portion of the scabbard body. If it fails to grip, then the body is produced from fiberglass. There is a prominent lip or ridge on the pommel cap which is not found on the originals. The diamond insigne on the pommel cap lacks the detail of the original as well. The use of rivets to retain the pommel cap was employed in the original specimens, and taken alone, cannot be considered a disqualifying feature. The second most prominent error is the larger size of the RZM mark in the circle. Original specimens had the circle contained totally in the flat surface of the blade whereas the reproduction is almost one-third larger, and extends over the flat surface towards the blade edges. The most common trademark used in the reproduction specimens is the RZM M7/36 (E & F Hörster). Later specimens went to the metal base scabbard and improved in the general overall appearance. A Marine Hitler Youth Leader version has been produced by using a blue leather scabbard covering and gold plating the metal portions of the scabbard and hilt. A presentation pattern has been produced, but this will be addressed in the following chapter.

NPEA STUDENT'S DAGGER

The blade is the main identifying feature of the reproductions in that it is hollow ground with a dull finish rather than the bright plated finish. Almost all specimens observed bear the "Karl Burgsmuller, Charlottenburg 5, Berlin" markings on the reverse of the blade. This distributor's mark was used on the original specimens and cannot be considered disqualifying in itself. The "Berlin" portion of the mark has been eliminated from current specimens. Early patterns were produced with a sharp ridge running down the center of the blade, but this was later corrected. The metal scabbard was modified from the SA, and can be detected by lightly scratching the scabbard tip. If the tip shows through in a gold color, this is the brass metal used in making the modification rather than the correct grey-colored steel. Crossguards were taken from original SA stock, and lack any NPEA designation or numbers as usually found on the originals. The grip was constructed from brown-dyed wood, with a aluminum eagle sometimes inset on the obverse. Later, grips were constructed from stained walnut, and are excessive in their size, giving an uncomfortable feeling when gripped in the hand. Early specimens used the black painted SA-style scabbard, but this was later corrected to the NPEA "bayonet" type scabbard, finished in olive drab or brown paint. An olive-grey stain has been used to finish the metal scabbard body of the most current specimens. The error of the two screws placed on the sides of the scabbard throat has been corrected by using a single screw centered just above the bayonet lug. The blade has been polished to take on a brighter finish as well, but is still hollow ground, and lacks plating.

It can only be assumed that a NPEA Leader's version has been or will be reproduced as well, but no specimen has yet been encountered.

LUFTSPORT-VERBAND DAGGER

Because so few original specimens were available from which to draw exact details, reproductions of the chained Luftsports dagger were more than difficult to achieve. The resultant product bears evidence of the poor workmanship which went into the making of the Luftsports dagger. The entire grip assembly was made from original 1934 Luftwaffe Dagger parts. The scabbard was constructed from an original 1934 Luftwaffe Dagger, elongated, and a Luftwaffe sword tip added. The blade appears to have been cut down from a Luftwaffe sword. The blade was of near proportions, and without a trademark. Since the blade was narrowed in its dimensions, the fit into the scabbard was too loose. In overall appearance, the resultant efforts to copy the Luftsports Dagger simply resulted in a 1934 Luftwaffe Dagger with elongated proportions.

DIPLOMATIC/GOVERNMENT OFFICIAL'S DAGGERS

This is another series of daggers which has undergone considerable transition since 1964, when it was first reproduced. The copies range from poor to excellent. The reason that the two daggers (Diplomatic/ Government Officials) have been grouped together is that comments for one pertain to the other, as the only distinguishing difference between the two is the direction which the eagle's head faces on the crossguard. The early specimens had a number of identifying features, most all of which have undergone correction. The tang button initially had a recess (to accommodate a screwdriver) instead of the two holes to accommodate a spanner wrench, but this was quickly rectified. The grip was constructed of translucent white plastic simulating mother-of-pearl, and were rather thick in their width. Looking at the grip from a frontal view, it gives a "potbellied" appearance rather than being streamlined. The outstretched wing tips of the crossguard eagle was filed straight rather than being upturned. In a few of the earlier examples, the chest of the crossguard eagle also gave a "potbellied" appearance rather than coming to a distinct "V". The blades used were normally original Army blades, and simply added to the original appearance. Obviously, the trademark would differ with each original blade used. Original Army scabbards were also used in the earlier patterns, but with the suspension bands sweated off, and Navy suspension bands braised on. This gave two immediate clues that the scabbard was altered, since the oak leaves ran away from the suspension ring rather than towards it. Also since the suspension bands

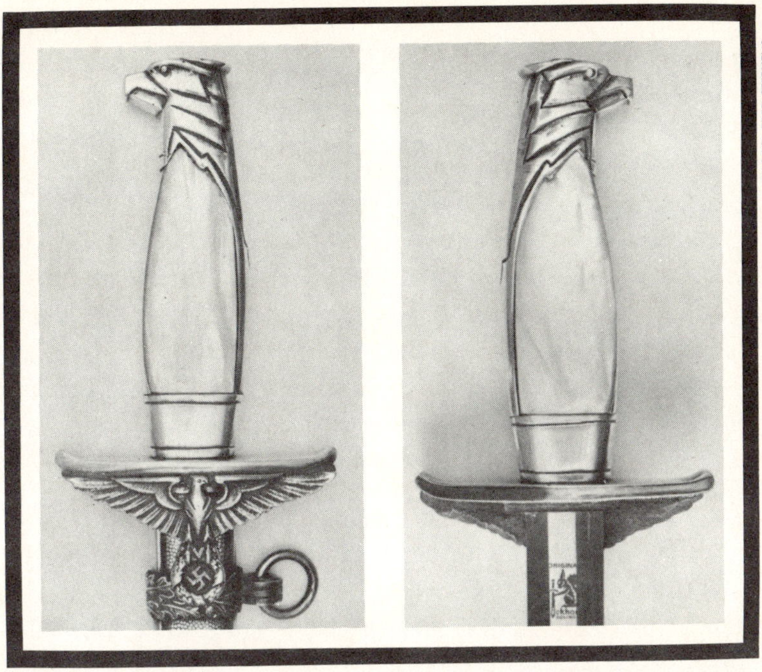

THIS RATHER GOOD REPRODUCTION OF THE DIPLOMATIC OFFICIAL'S DAGGER HAS A NUMBER OF SUBTLE ERRORS. THE "POTBELLIED" APPEARANCE OF THE GRIP FIRST MEETS THE EYE, FOLLOWED BY THE TRANSLUCENT LOOK OF THE PLASTIC GRIP PLATES, WHICH SIMULATE MOTHER-OF-PEARL. THE LACK OF THE UPTURNED WINGS OF THE EAGLE CROSSGUARD IN ITSELF SHOULD NOT CAUSE THE PIECE TO BE IDENTIFIED AS A REPRODUCTION.

were constructed of brass, a magnet would not attract as with the original. More recent specimens made in Germany have been manufactured in total with exacting detail. Silver specimens have been gold-plated, but in most cases the plating lacks consistency and wears easily.

RED CROSS LEADER'S DAGGER

This dagger has presented a great deal of difficulty for the collector to identify as a reproduction, since most of the parts used were taken from original stocks. These parts include the scabbard, and metal portions of the grip. The newly cast plastic grip lacked the distinct detail, and was near white in color rather than a creamy yellow. In addition, there was a distinct mold line down each side of the grip. The blade was hollow ground, and totally lacked the mirror-like finish which was characteristic of the originals. As with the originals, the reproductions had no trademark or other blade markings.

TeNo LEADER'S DAGGER

This specimen is one of the newest of the reproductions and with the exception of some casting problems, the piece is well executed and nearly exact in weight and configuration. The pommel and crossguard show a number of small pits which came from poor casting. Many have been polished out but many remain. The detail in the cogwheel lacks sharpness and the detail and configuration of the crossguard eagle is cream-colored rather than yellow, and lacks definition in the ridges. The blade is unplated steel, and lacks the mirror-like finish of the original. The Eickhorn trademark is used, but with almost twice the normal separation from the TeNo eagle. The serial number (specimen piece Nr. 1643) was etched and is larger than normal. The serial number is stamped with uneven spacing on the top of the scabbard throat. The scabbard suspension bands show traces of unfinished metal remaining in the grooves, which has been plated over. No effort was made to burnish the scabbard as with the originals, and therefore are of a bright, silver-like finish. When all the metal parts were polished, they took on a bright luster. Given time and an accelerated aging process, this specimen would be extremely difficult to identify as a reproduction.

LAND CUSTOMS DAGGER

The first reaction of a novice collector is to think that the crossguard and pommel are rather crude when viewing a Land Customs Dagger, but this was the case with the originals. Therefore, lack of detail in these two parts should not be considered. One noticeable flaw when looking at the grip is that the grip wire was retained by means of a wooden peg. In fact, in the very earliest specimens, the grooves and wire actually run in the wrong direction, i. e. from lower right to upper left. The leather wrap on the grip and scabbard is absolutely new, but of the same color and texture of the original. In some cases, however, a line was impressed down each side on the front of the scabbard. The scabbard tip in most cases was considerably larger than normal and lacked the tapered configuration of the original. In some specimens, there was no effort to copy the throat lip, thus resulting in an unfinished appearance.

SEA CUSTOMS DAGGER

The same comments apply for this piece as stated above. Only the early specimens were modified from the white aluminum Land Customs Dagger by means of a light gold wash. Later copies were constructed of solid brass and were excellent in detail. The scabbard tip ranged from the tapered to the untapered versions, and the blue leather was with and without the mold lines. While original Army blades were either initially used or exchanged with the reproduction, later copies used reproduction blades which were hollow ground, and were etched or stamped with a wide variety of trademarks.

SS(BLADE ONLY)

At least six specimens have been observed which were manufactured in 1972. In some cases, these continued to show the sharp center ridge, but in others, the ridge was rounded. While the blades had some semblance of a luster, they still retained the dull finish characteristic of the hollow ground blades. The trademark used was the SS sigrunen in a double circle over 807/36, and above the RZM in a double circle. The weight appeared to be a bit heavier than an original blade.

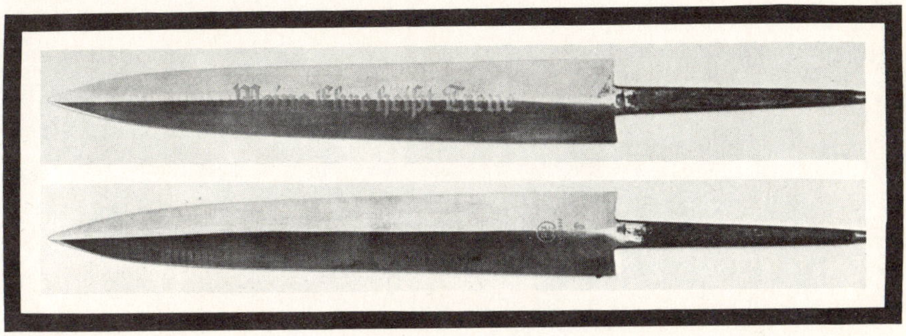

PHOTO: ANDY WALKER

NOTE THE SHARP RIDGE RUNNING DOWN THE CENTER OF THIS REPRODUCTION SS BLADE INSTEAD OF THE ROUNDED AND LESS DISTINCT HIGH POINT OF AN ORIGINAL. THE SQUIRREL TRADEMARK, BELOW THE "RZM" MARK, IS PROBABLY INTENDED TO INFER MANUFACTURE BY THE EICKHORN FIRM, BUT NO SUCH TRADEMARK WAS EVER USED.

NAVY PRESENTATION DAGGER

The firm of F. W. Höller offered a Navy presentation dagger complete with presentation case starting in the mid-1960's, and continues to offer these pieces for sale today. They are exact reproductions of patterns which they produced and sold during the war years. Early specimens were produced from remaining stocks, but as these diminished, the entire dagger was reproduced. In comparison, it is the most finely made dagger available, and is equal in quality to the original in almost every respect. For want of some distinct feature that might be hit upon to identify these, I must resort to minute detail. The genuine

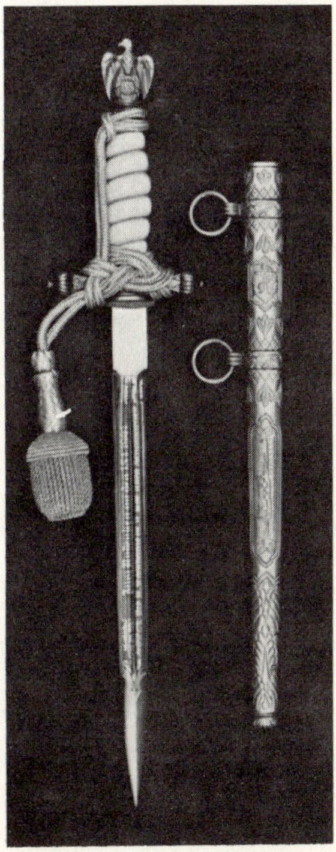

ALBERT GOOSENS

THIS NAVY PRESENTATION DAGGER WAS ALLEGEDLY AWARDED TO OBERLEUTNANT GÜNTER PRIEN IN 1941. SUFFICE IT TO SAY THAT PRIEN'S FIRST NAME WAS GÜNTHER, AND HE WAS A KORVETTEKAPITÄN FROM AT LEAST TWO YEARS EARLIER.

ivory grip lacks the yellowing of age or the grain detail of ivory used during the war period. The blade was etched with a wide variety of patterns and presentations, with some portions of the presentation sometimes being plated in gold. What separates this piece from the standard reproduction is that the blade is finely plated and finished. Instead of going into detail describing this particular piece, see the illustrations of example pieces. The presentation case lacks all detail of the original. The lining is loose fitting, blue satin both top and bottom. The dagger lies in the case by means of two ramps rather than being inset as with the original. The brass retaining fasteners are identical to the original.

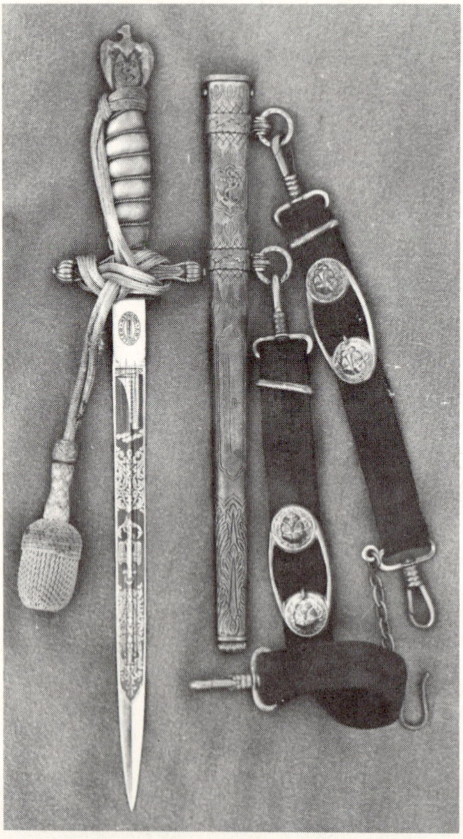

PHOTO: MARC BROWN/HICKOX COLLECTION

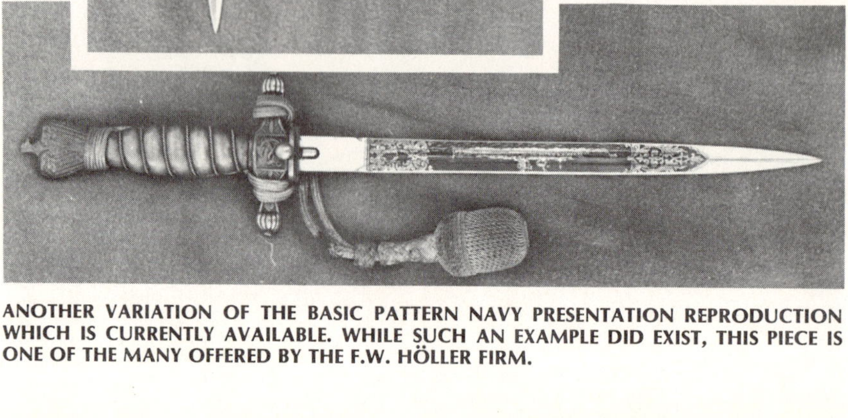

ANOTHER VARIATION OF THE BASIC PATTERN NAVY PRESENTATION REPRODUCTION WHICH IS CURRENTLY AVAILABLE. WHILE SUCH AN EXAMPLE DID EXIST, THIS PIECE IS ONE OF THE MANY OFFERED BY THE F.W. HÖLLER FIRM.

CONDOR LEGION MAIN GAUCHE

Extensive research has been conducted to determine if the Condor Legion ever had a recognized blade to reflect its exploits in the Spanish Civil War (1936-1939), and these efforts have resulted in the assumption that no such "dagger" was ever authorized. However, considering the pattern of the "Main Gauche", it is highly probable that souvenir letter openers were produced on an individual order basis, and affixed with the dedication shield as prescribed by the purchaser. One such possible original souvenir blade exists in the collection of Mr. Norman C. Heilman. In 1965, approximately ten copies of this piece were reproduced from photographs. They were initially sold as copies at the price of ten dollars each, with no intent to defraud. As the pieces changed hands, however, the copies made their transition to "originals" in some cases. The copies had an overall length of 10 1/2 inches with a blade length of 7 3/4 inches, and were constructed of steel. All hilt fittings are brass,

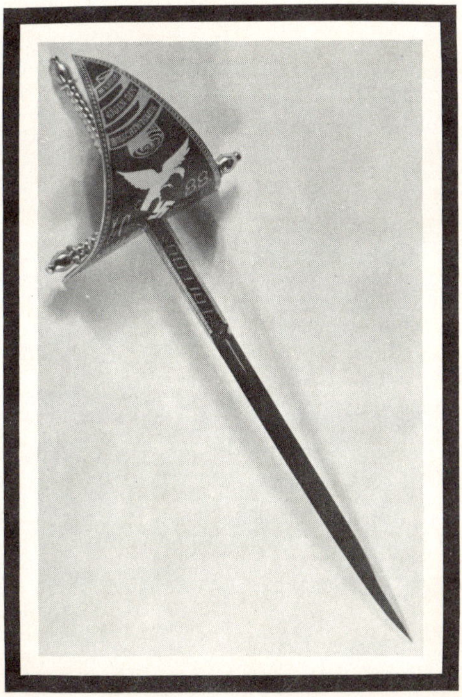

ILLUSTRATED IS A POSTWAR COPY OF THE "CONDOR LEGION MAIN GAUCHE", ALLEGEDLY MADE IN SPAIN FOR THE MEMBERS OF THE CONDOR LEGION. NO SUCH OFFICIAL BLADE WAS EVER INTENDED FOR THE LEGION MEMBERS.

with a crossguard span of 3 inches. An insignia plate, affixed to the basket by two prongs, bore the dedication "Im Kampf gegen den Bolschewismus L.C. 88. Spanien 1936-38". The blade bore the name "Toledo", and is characteristic of souvenir letter openers still produced in Toledo, Spain today.

NAVAL ASSAULT DAGGER

Subsequent research has disclosed that this dagger did in fact exist as an authorized Naval variation. Most of those encountered in collections today, however, are probable reproductions. The dagger design was the sole patent of the Puma Werks, Solingen, and was illustrated in that firm's 1940 sales catalogue. The designation afforded the dagger by that firm was "Deutscher Marineoffizierdolch Nr. 8", and no reference was made of it being solely designated for Naval Assault Forces. Since the dagger was only produced by a single firm, the variations in the design details of the crossguard cannot be attributed to "manufacturers' variations". The most significant difference between the original pattern and those of postwar production is the eagle insignia which is the central design feature of the crossguard. Specifically, the shoulders of the original design eagle are rounded, while reproductions overextend the shoulder in a straight line. In some cases, standard pattern Navy pommels were used instead of the specially designed pommel which is unique to the Puma dagger. Any dagger of this design having a blade bearing the trademark of any other firm than Puma, should be suspect.

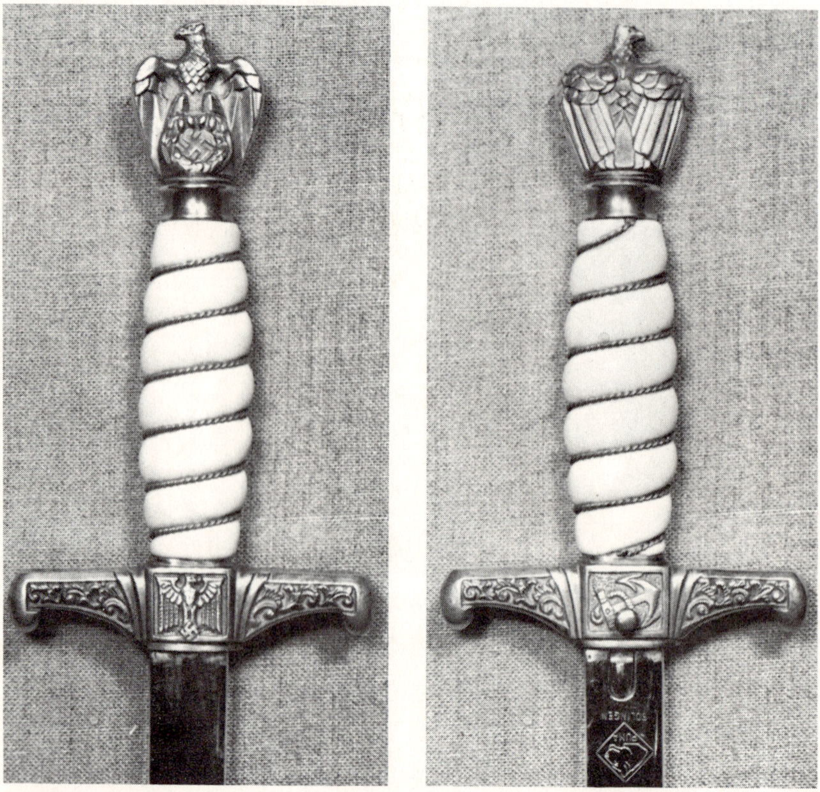

COMPARE THE DAGGER ABOVE WITH THE ORIGINAL PUMA DESIGN SHOWN ON PAGE 121. NOTICEABLE DIFFERENCES EXIST IN THE SCROLL WORK AND DESIGN OF THE WEHRMACHT EAGLE. NOTE THE STANDARD PATTERN POMMEL USED.

FIRE DEPARTMENT OFFICIAL'S DAGGER

This pattern dagger is currently being produced by the Eickhorn firm and is actively in use as a dagger sidearm. Because it is being made by the original firm using the same dies, only its newness gives any hint of date of manufacture. While specimens are found with both the nickel and gold finish, Eickhorn stated that the firm did not produce a Fire Department Official's Dagger with gold finish prior to 1945, but that current production pieces can be obtained with the gold finish. It should be emphasised that the dagger being produced today is not considered a reproduction with any intent to deceive, but is a production item for current consumption. Unfortunately, there are few ways to distinguish the new pieces from the earlier production blades. As the current pieces are aged, it will be difficult if not impossible to tell them from the originals.

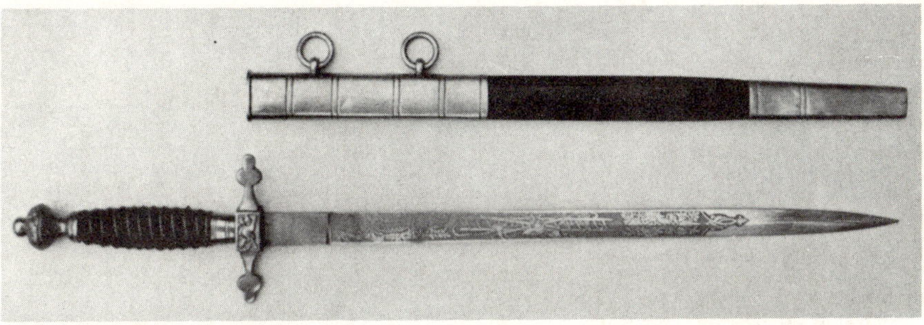

THE CURRENT PRODUCTION FIRE DEPARTMENT OFFICIAL'S DAGGER IS IDENTICAL TO THAT PRODUCED DURING THE NAZI PERIOD, AND ONCE AGED, WILL BE ALMOST IMPOSSIBLE TO DETECT FROM AN ORIGINAL.

MARKSMANSHIP ASSOCIATION OFFICIAL'S DAGGER

The same comments apply for this piece as with the Fire Department Official's Dagger. Again, when aged, detection will be virtually impossible. Obviously, the Nazi-period Marksmanship Association insigne is added to the shell guard after sale by the manufacturer.

RAD LEADER'S DAGGER (MINIATURE)

This is an excellent example where a few isolated original parts were located and transformed into a complete specimen. The hilt is made of brass and is lightly plated. The white plastic handle is affixed by means of a pin through the face of the grip. This is the main distinguishing feature as the original had no such pin. Unlike the original, the stamped steel blade is unetched, and is not compatible with the fit of the scabbard. The only specimens (original) which have been encountered have been produced by the Alcoso firm, and have a dark yellow or orange grip.

ARMY OFFICER'S DAGGER WITH SPECIAL PURCHASE BLADE

Normally, this particular series cannot be classified as a true reproduction, since in most cases, it is the blade pattern that becomes the point of interest rather than the dagger. Since original pieces were used in most cases, no single manufacturer can be cited for initial identification. While manufacturers' quality and age of the pieces have varied, the template used to achieve the etching has been consistent in almost every case. The basic pattern is a floral design with the words, "Unserem Kameraden der Regimentstab". The bar lines at each end of the pattern are broken rather than solid, probably due to the photo-etching process. In one example, the entire pattern was off-set on the face of the blade. Unlike original specimens, the blades are not chrome plated and then etched, but are etched directly on the unplated steel blade. Thus, the blades tend to take on a dull, rather than bright appearance. Early examples were etched only on a single side with names and ranks engraved on the reverse (examples: Oberst Beck, Hauptmann Zell, Oberleutnant Huser) and filled with a gold color. Later patterns were made with a double-etched design, however, with the reverse pattern usually including the Army (and sometimes Luftwaffe) eagle.

SS DAMASCUS LETTER OPENER—P. MULLER

At least one dozen of these damascus letter openers were manufactured at an estimated cost of $20 each, just prior to their import into the United States in 1964. They were alleged to be originally made at the Dachau Concentration Camp metal forge, to be given as Christmas gifts by Reichsführer-SS Heinrich Himmler. The only thing that can be said about the piece in terms of originality is that they were made by Paul Müller, but after the war. The specimen is made from a single band damascus steel (Turkish). The obverse bears the raised gold signature "H. Himmler", while the reverse shows the SS runes in a circle and "Echt Damast P. Müller", also raised and gold-plated.

There are other pieces, and probably more to come which have been reproduced, but either insufficient information or tell-tale features exist to provide a detailed description. These are but a few which fall into this category: Fire Department Dress Axe; Navy Miniature (especially those bearing the Prien inscription); 1934 Pattern Luftwaffe; SS Miniature (usually a single metal piece construction without scabbard); Regulation Hunting Association Sidearm (current production piece for modern consumption, but with a reproduced Association insigne to replace the current DLV insigne); Railway Protection Leader; Canal and Waterways Protection Police Leader (nickel and gold patterns); standard dress sidearms with etched blades (again, current production pieces with war-time templates used to made the transition); Army Field Marshal's Dagger, etc., etc.

As long as the value of daggers continues to rise, there will always be reproductions and alterations. While there is little that we can do about the latter, there are certain legal safeguards that exist to save us from reproductions manufactured in another country. U. S. Customs regulations require that the country of origin be die-sunk on the imported piece providing the item is not of W. W. II or earlier manufacture. That same regulation goes on to state that any person intentionally concealing that mark by defacement, removal or other obliteration, commits an offense punishable, upon conviction, by fine or imprisonment, or both.

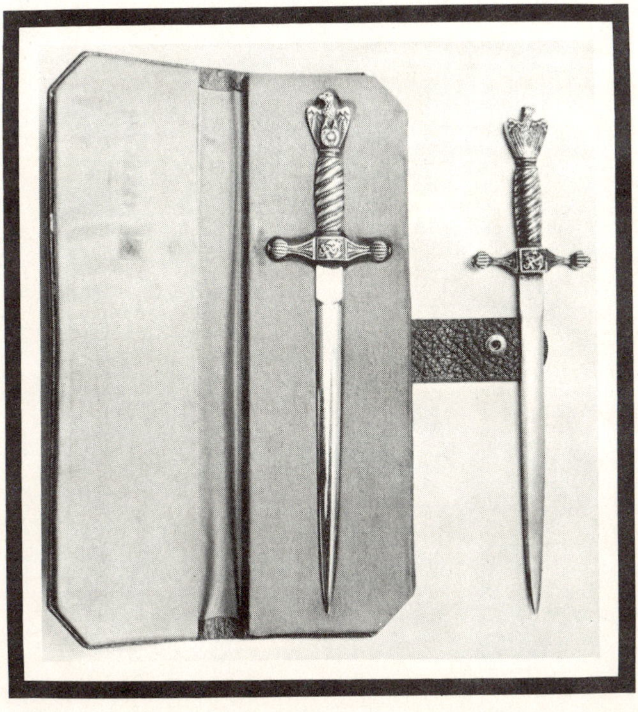

CURRENT PRODUCTION LETTER OPENER BEARING THE SAME BASIC DESIGN AS THE 1938 PATTERN NAVY DAGGER. NOTE THE TOTAL LACK OF SWASTIKA ON THE POMMEL EAGLE.

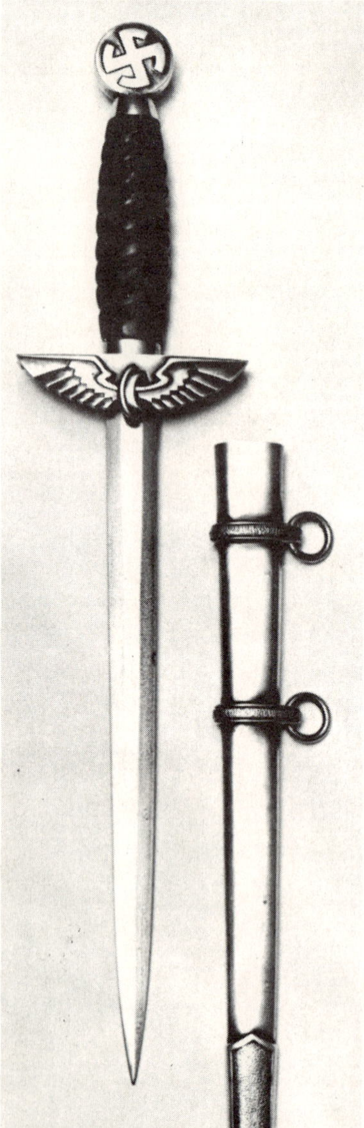

PHOTO: ANDY SOUTHARD/COL. C. J. STEWART

MODEL 1941 PATTERN RAILWAY LEADER'S DAGGER AS DESCRIBED IN THE MARCH 1, 1941 ISSUE OF "UNIFORMEN-MARKT". WHILE IT IS APPARENT THAT A VERY FEW SPECIMENS WERE MANUFACTURED AS PROTOTYPES, NO KNOWN ORIGINALS HAVE YET APPEARED. AT LEAST ONE COPY IS KNOWN TO HAVE BEEN REPRODUCED.

PHOTO: FRED STEPHENS

A SIMPLE PERSONALIZED ADDITION TO AN ORIGINAL BLADE WILL ADD APPROXIMATELY $100 TO THE OVERALL VALUE. THIS PIECE PROBABLY WOULD HAVE GONE UNDETECTED HAD IT NOT BEEN ACKNOWLEDGED AS AN ALTERATION AT TIME OF INSPECTION.

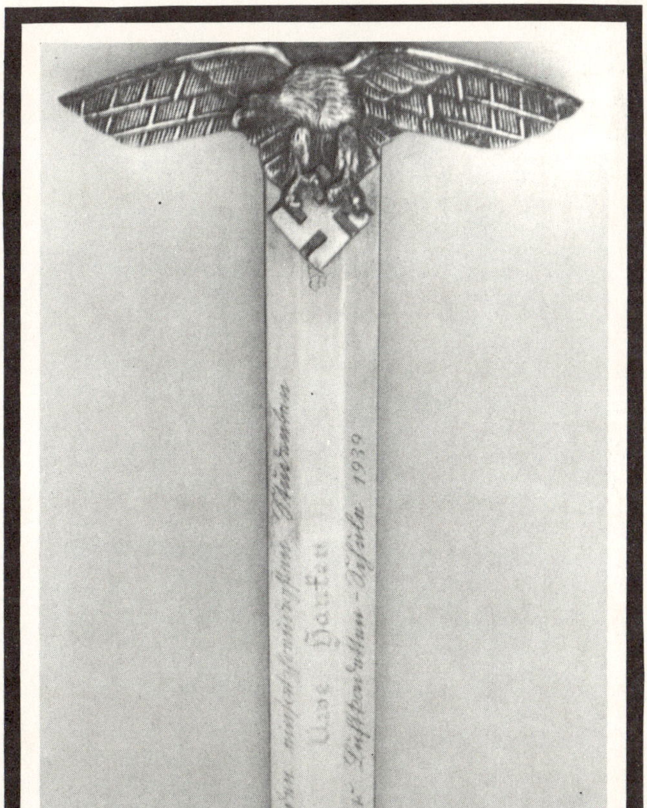

PHOTO: FRED STEPHENS

AN ORIGINAL 1937 PATTERN LUFTWAFFE DAGGER, WHICH HAS HAD A BOGUS DEDICATION OR PRESENTATION EXPERTLY ADDED TO THE BLADE, IS VERY DIFFICULT IF NOT IMPOSSIBLE TO DETECT. CONSIDERABLE VALUE IS ADDED TO A GOOD BASIC PIECE BY SUCH AN ALTERATION.

MAJ. TOM JOHNSON

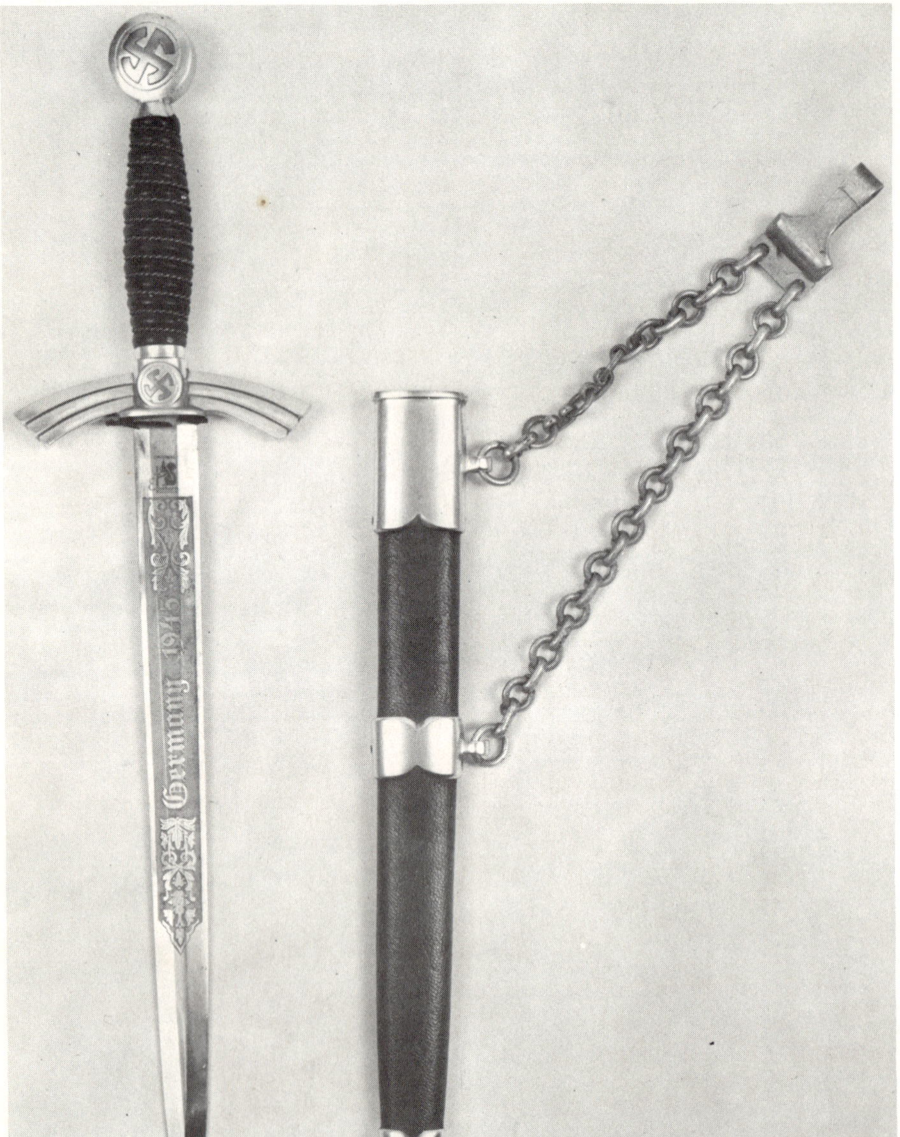

NOT CONSIDERED A REPRODUCTION PER SE, THIS 1934 PATTERN LUFTWAFFE DAGGER WAS TAKEN AS A WAR TROPHY FROM ON-HAND STOCK AT THE END OF THE WAR. THE BLADE WAS ETCHED AT THE REQUEST OF THE AMERICANS, PROBABLY TO BE GIVEN AS A PRIZE. THE REVERSE OF THE BLADE IS ETCHED WITH THE WORDS, "FIELD DAY".

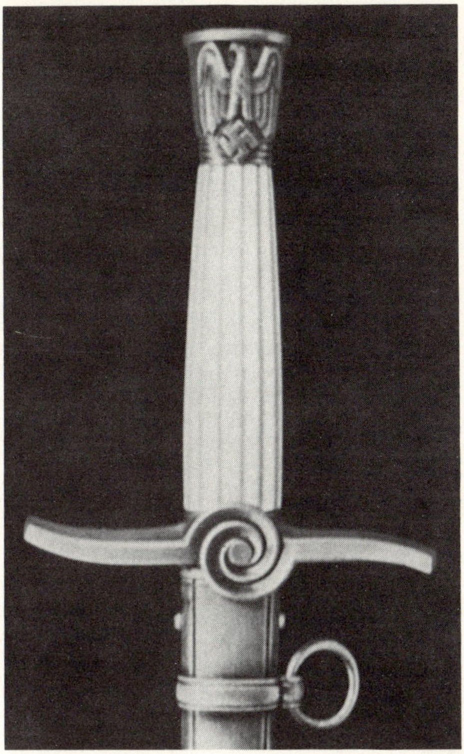

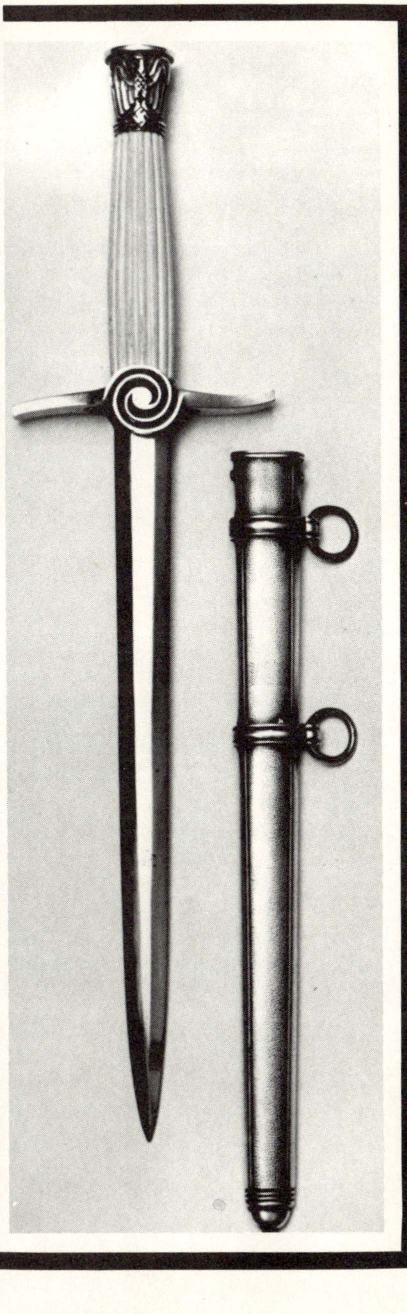

PHOTO: ANDY SOUTHARD/COL. C. J. STEWART

THE ARMY FIELD MARSHAL'S DAGGER CANNOT TECHNICALLY BE CONSIDERED A REPRODUCTION SINCE AN ORIGINAL WAS NEVER MANUFACTURED. THE CURRENT DESIGN WAS TAKEN FROM AN ORIGINAL DRAWING OF PROPOSED SIDEARMS BY THE INDUSTRY AND TRADE UNION OF SOLINGEN, WHICH WAS ILLUSTRATED IN "BLANKE WAFFEN". THE COLORED RENDERING BORE THE INSCRIPTION "ARMY OFFICER'S DAGGER". EVEN THOUGH THERE ARE SOME DIFFERENCES, THE CURRENT PRODUCTION PIECE BEARS A VERY CLOSE RESEMBLENCE TO THE WAR-TIME PROPOSAL.

This alleged SS fighting knife has no origin whatsoever during the period prior to May 1945, but is in fact of current manufacture. It is believed to be produced for the British Engineer Corps for use as a utility knife. An extendable pentograph, the piece in its closed state measurers 14. 5cm long with a width of .75cm when the blade is fully concealed. In its open, extended state, the piece measures 23. 9cm long, the blade being 9. 5cm in length. The blade is double edged, and fullered on one side only. The outstanding feature of the knife are its extensive markings... ...the early Eickhorn trademark accompanied by the SS proofs on the blade, the SS eagle on the securing clip, the RZM M7/66/34/SS code on one side of the "grip" cover, and "SS Streifendienst" on the other. The piece is of all metal construction and rather cheaply made. Many pieces examined have been artificially aged and pitted.

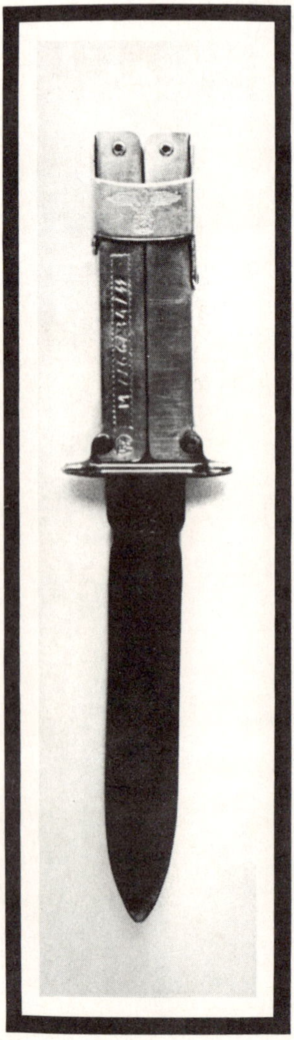

PHOTO: N. C. HEILMAN

ONE OF A SERIES OF WHAT IS BELIEVED TO BE PURE FANTASY PIECES. THE ORIGINAL WATER COLORS ARE EXTREMELY WELL EXECUTED, AND SHOW A GREAT DEAL OF IMAGINATION, BUT IT IS BELIEVED THAT THEY ARE THE WORK OF AN AMERICAN, CIRCA THE MID-1950's.

The following listed firms are still in business today, and were manufacturers of daggers, swords and other assorted sidearms during the period 1933-1945:

Carl Eickhorn
Alcoso-Werk GMBH
Paul Seilheimer
F.W. Höller
Carl Julius Krebs
WKC - Stahl- und Metallwarenfabrik GMBH
Puma-Werk Lauterjung & Sohn
Kuno Ritter
Robert Klaas
E & F Hörster
Wilh. Wagner
Eugen Witte
Solinger Metallwarenfabrik Stucker & Co. (SMF)
Lauterjung & Co. Tiger-Stahlwarenfabrik
Scharff, Kober & Co.
Gottlieb Hammesfahr
C. Eppenstein Sohne

ROBERT KÜRTEN, SOLINGEN'S LEADING DAMASCUS CRAFTSMAN, USIN
MODERN EQUIPMENT IN THE PRODUCTION OF A DAMASCUS BLADE.

The New Damascus Art

When Paul Müller, last remaining damascus artisan in Germany, died in Solingen in 1971 at the age of 83, it looked as if the art of damascus blade creation had come to an end. Granted, Müller had made every effort during the war years to train assistants to learn the trade of damascus smithy, but his three assistants were drafted into military service in 1943, leaving him alone once again to meet the constant demands for damascus production. The men that Müller had trained did not return to the forge to test their newly acquired skills, and no information is available concerning their fate on the battlefield.

Müller was still performing his trade six months prior to his death, but his hand was shaky with age, and he had lost the keen sight necessary in hammering a straight edge by eye. When Müller died, his craft did not die with him, however. Realizing that his age and eventual death would bring an end to a long tradition of damascus production, Müller sought and found someone skilled enough to be trained as a damascus smith. The man who was to receive the secrets of the damascus trade was Robert Kürten. With his selection, the complex and strenuous training began in an effort to provide him with the necessary skills in a few short months, which had taken Müller years to gain mastery of.

Robert Kürten was fifteen years old when he became an apprentice in the blade trade in 1940. His work in Solingen was interrupted when he was drafted into military service to serve with the Army. At the end of the war, Kürten returned to Solingen to continue in his craft. It was not until 1970, that he was selected to become the pupil of Paul Müller. He considered himself extremely fortunate to have such a master for a teacher. Müller was self-taught in the trade, and Kürten was not subjected to the rigors of similar trial and error. Even in his old age, Müller was able to guide and mold his apt student. Every stroke of the hammer, every bellow of the forge came under Müller's watchful eye. A more than capable student, Kürten learned his trade in six months. The old master, content in the fact that he had given the world a competent artisan to take his place, died without actually seeing Kürten being accepted into the elite guild. Robert Kürten underwent his guild tests, however, and on October 4, 1971, was proclaimed a damascus master by the "Handwerkskammer" in Düsseldorf, after receiving a special consideration by the Düsseldorf head of government.

ROBERT KÜRTEN

Kürten continued in the hire of the Eickhorn firm as a consultant master, following in the foot steps of such previous Eickhorn damascus masters as Paul Dingler, Karl Wester, and the father and son team of Alexander and Paul Müller. Just prior to Müller's death, Kürten acquired the sole rights to use Müller's trademarks on blades produced by Kürten. It would seem fitting that the Kürten blades should bear the Müller mark as they bore the tell-tale techniques of Müller himself. Kürten continues to this day to use the Müller trademarks rather than his own. The markings include the tang mark used to identify the master craftsman. At least four patterns of Müller identification stamps are used to finish off the blade work.

DAMASCUS PRODUCTION

While Robert Kürten is extremely proud of his creations, and rightfully so, he is hesitent to discuss the actual techniques he employs to achieve the finished results that are so greatly sought after. Watching the master at his trade was sufficient to determine that his skill was the result of long and arduous training.

He moves in quick, precise steps. Using his eye as his main tool, he transfers the image he sees in his mind's eye to his skilled hands, and the creation begins to take shape on the forge. Mr. Kürten's secret is in the manner in which he builds up the steel. Distaining false damascus, Kürten prefers to work with genuine Turkish damascus. Starting with only the best quality steel, he is able to build the blade in a manner that no press can duplicate. The blade begins with the hot damascus steel rods, with the number of rods necessary depending on the length, width and thickness desired. Placing the square rods, normally three, side by side, he begins the process of heating and hammering until they become a single unit. He hammers out the desired basic form, and continues the process until he has given it the right degree of strength. He then tapers the edges, and forms the finished shape. As this stage of his work nears completion, Kürten's hand becomes that of a true artisan, straightening an edge by eye and providing just the right degree of temper to the blade that only years of painstaking effort and experience is capable of achieving.

Once the blade form is complete to satisfaction, it is polished to the point that it looks like any other standard steel blade, void of any damascus lines or pattern. The trademark, motto and dedication as specified by the purchaser are then applied to the blade through an acid process. It is not until a series of acid baths that the damascus pattern is drawn out, the acid eating away at the softer steel at a faster rate than it dissolves the harder steel. With the desired pattern achieved, the final touches are put on at the anvil. A glass-fibered stick is applied to the blade to bring the damascus pattern to a bright luster, and remove any small blemishes. A clear lacquer is then applied to the blade to provide protection. The blade is then fitted to the desired hilt, and the process is complete. Production depends on the size and width of the blade, but it normally takes between 12 and 40 hours to complete, depending on the degree of difficulty. There is no wonder that Robert Kürten's creations command high values.

Other Solingen craftsmen not having the talent or training that Kürten possesses, resort to producing artificial damascus to satisfy the demand, but it is a poor substitute. By means of a die or metal stamp, they hammer the desired pattern into the blade. The patterns... rose, mosaic, ribbon maidenhair and peacock feather... are quite beautiful to look at, but do not require the artisan's touch necessary to produce the genuine damascus blades. Another process involving the use of a design template and acid bath is also used to achieve the artificial damascus patterns. While Mr. Kürten produces the artificial damascus upon request, he looks upon them with disdain when comparing them to his own handiwork. He rarely resorts to producing an artificial damascus pattern by means of acid etching. He leaves such techniques to the mass production market. The tangs of his blades legitimately bear the designation "Echt Damast" (genuine damascus) or "Damast".

Robert Kürten does not intend to let the secret of his trade end with him. He currently has as his apprentices, his son, son-in-law, and when age allows, his grandson.

The Solingen craft guild became so concerned during the war that something might happen to cause the art to die that they produced a full-length film of the entire process, and filed it in their archives. Because of Kürten, they have never had to resort to the use of this film as a training aid.

DAMASCUS REPRODUCTIONS

During the 1960's, the principal producer of reproduction damascus blades was the old master himself, Paul Müller. The blades failed to measure up to his war-time creations in that they did not achieve the same degree of thickness, weight, finish or skill. The differences can probably be attributed to Müller's advanced age, and the lack of detailed quality inspection which the earlier pieces were subjected to.

Probably Müller's most noteable reproduction is the Himmler damascus letter opener, allegedly produced for Reichsführer-SS Heinrigh Himmler by Müller at the Dochau Concentration Camp metal forge. These were to have been given by Himmler to selected recipients at Christmas time as the story goes. Müller produced at least a dozen of these in the early 1960's and their production continues upon request, produced this time by Müller's capable student, Robert Kürten.

Kürten is now master of the reproductions, capable of producing any pattern. . . old or newly created. . . limited only to the purchaser's ability to pay. Here are but a few of the pieces that are turned out for today's reproduction market:

1. SS Honor blade of genuine damascus steel. The obverse bears the standard SS motto, "Meine Ehre heisst Treue", preceded and followed by oakleaves. The motto and oakleaves are raised from the blade level, and lightly gold-plated. The reverse bears the raised gold dedication, "Julius Schreck für höchste Verdienste", and the signature, "H. Himmler". Also shown is a wide selection of trademarks and the words, "Echt Damast", also raised and lightly gold-plated. The blade is then assembled with the reproduction Honor-type crossguards, a regulation grip, and fitted to a leather-covered scabbard having Honor-type mounts.
2. Model 1934 Pattern Luftwaffe blade of genuine damascus. The obverse shows a wide variation of names and ranks, each having a Luftwaffe eagle fore and aft. The reverse bears the raised gold dedication, "Für besondere Tapferkeit im Luftkampf", followed by the signature of Hermann Göring. A selection of trademarks can also be found on these pieces. The finished blade is mounted to expertly crafted fittings.
3. SA Honor, complete.
4. SS Honor, complete.

5. High quality SA Service Dagger with damascus blade, complete.
6. Navy Honor, complete.
7. Model 1936 Pattern SS.
8. Navy Officer's Sword with damascus blade.
9. SS Officer's Sword.
10. Justice Official's Sword...the hilt is solid brass, polished to a high luster, and unplated.
11. Forestry Official's with damascus blade. Some bear a raised gold dedication to include Göring's signature.

While this is only a partial listing of the pieces Mr. Kürten is able to provide, it readily becomes obvious that he is capable of producing any blade form to order. One has to see his handiwork to appreciate the beauty and craftsmanship which each projects.

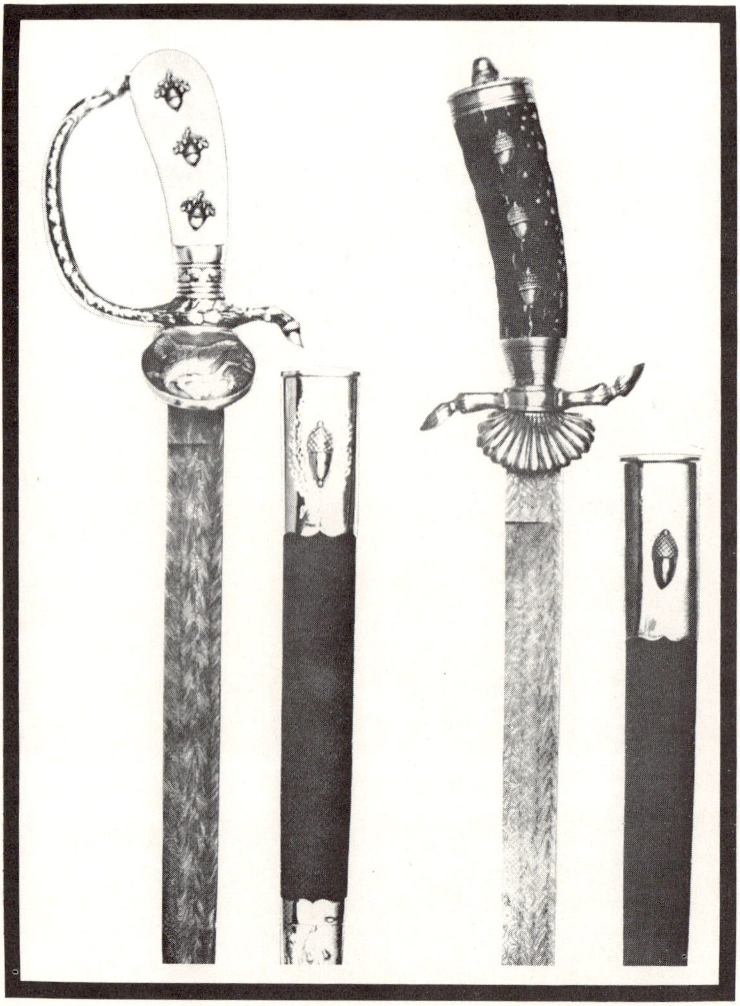

AN EXAMPLE OF ROBERT KÜRTEN'S FINE CRAFTSMANSHIP.

COMPLETE NAVY HONOR DAGGERS ARE PRODUCED TO ORDER. FOR EXAMPLE, THE BLADE SHOWN BEARS THE DATE 31 AUGUST 1942, AND HAS NO RELATIONSHIP TO THE DATES OF ACTUAL AWARD, BUT WAS PLACED THERE AT THE INSTRUCTIONS OF THE PURCHASER. WHILE THE WEYERSBERG TRADEMARK IS USED ON THIS SPECIMEN, ANY TRADEMARK IS READILY AVAILABLE. NOTE THE USE OF THE PM (PAUL MÜLLER) MARK ON THE UPPER TANG.

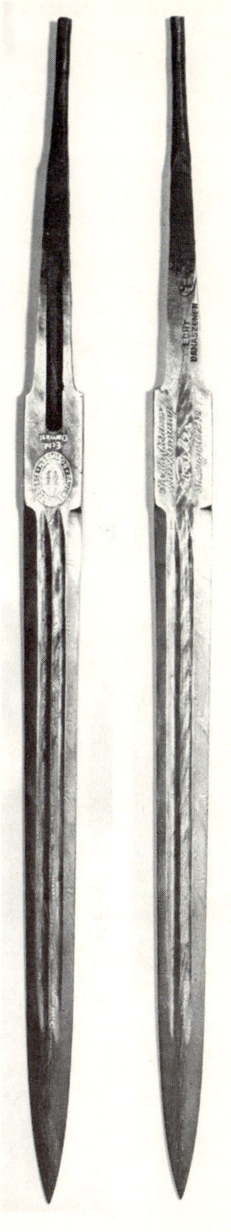

PHOTO: ANDY WALKER

HITLER YOUTH LEADER'S DAGGER WITH DAMASCUS PRESENTATION BLADE. ALL METAL FITTINGS ARE GOLD COLORED BRASS, AND ARE OF HEAVY CONSTRUCTION. THE CROSSGUARD IS ALMOST TWICE AS THICK AS THE STANDARD PATTERN.

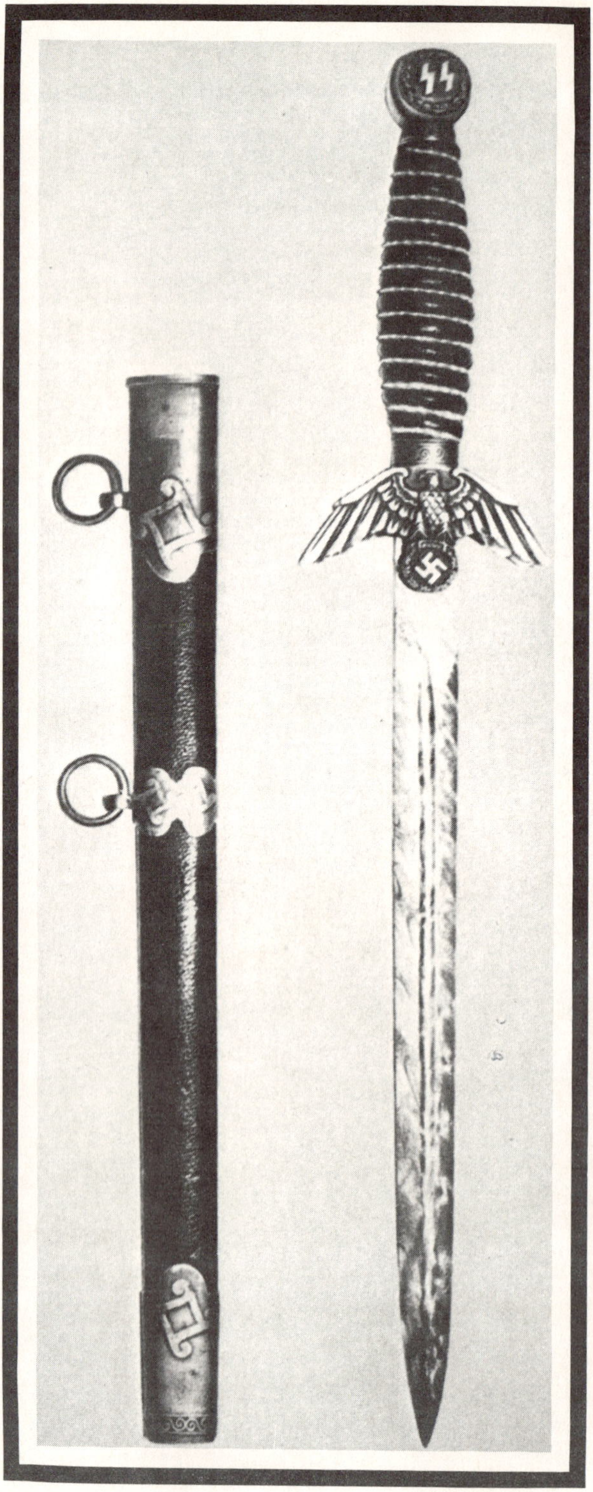

PHOTO: FRED STEPHENS

PHOTO: ANDY WALKER

A NEWLY DESIGNED SS PATTERN WAS PRODUCED USING THE EARLIER REPRODUCTION PATTERN AS A BASE. HILTS, LIKE BLADES, CAN NOW BE MADE TO MEET ANY BUYER'S REQUIREMENTS.

PHOTO: ANDY WALKER

Not all reproduction damascus blades are made by Mr. Kürten, for the old master, Paul Müller, produced his share before his death. Nor are all pieces produced by masters such as Müller and Kürten. The example shown here is not true damascus, but a simulated damascus pattern achieved through acid etching, and produced by an absolute novice as a test. Mr. Herman Ward, using original blades, produced ten of these artificial damascus blades using a tedious process of trial-and-error. Tracing the desired design, it was transferred to the blade by means of tar paper, followed by an acid bath. Some remained too long in the acid, some not long enough. However, a few achieved the desired results. This was not done in an attempt to deceive, but merely to test the process. This blade was provided courtesy of Mr. Ward.

Miscellaneous Daggers, Knives, etc.

The unattributed dagger above was manufactured by the E & F Hörster firm in Solingen. The grip is covered with shark skin over wood, and bound with a twist wire wrap. All hilt and scabbard fittings are made of brass. The date of production is about 1934. A similar pattern dagger with an etched blade is known to exist.

This dagger is an early pattern Eickhorn production piece, but unattributed as to organization or purpose. It was probably a special award dagger produced in late 1933. Hilt and scabbard fittings are high luster, gold plate, while the leather takes on a purple hue. No other specimen of this dagger is known to exist.

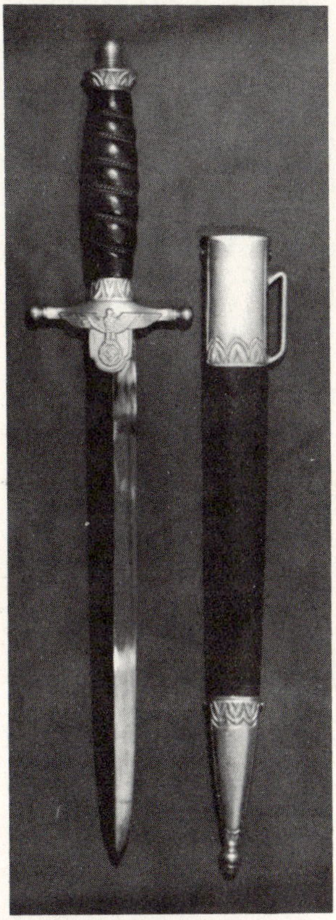

WAGNER-PIAR COLLECTION

HAUPTMANN UEBE LOOKS ON AS A MEMBER OF HIS CREW TAKES A DRINK TO CELECTRATE HIS 60th MISSION. THE ENLISTED CREW MEMBER WEARS A NON-REGULATION HUNTING KNIFE, PROBABLY PURCHASED IN FINLAND WHERE THE UNIT WAS BASED.

HAUPTMANN UEBE, HERE BEING CONGRATULATED BY GENERAL JODL, ALSO WEARS A NON-REGULATION SIDEARM. THE PURCHASE AND WEAR OF "SURVIVAL" KNIVES ON A PERSONAL BASIS WAS QUITE COMMON.

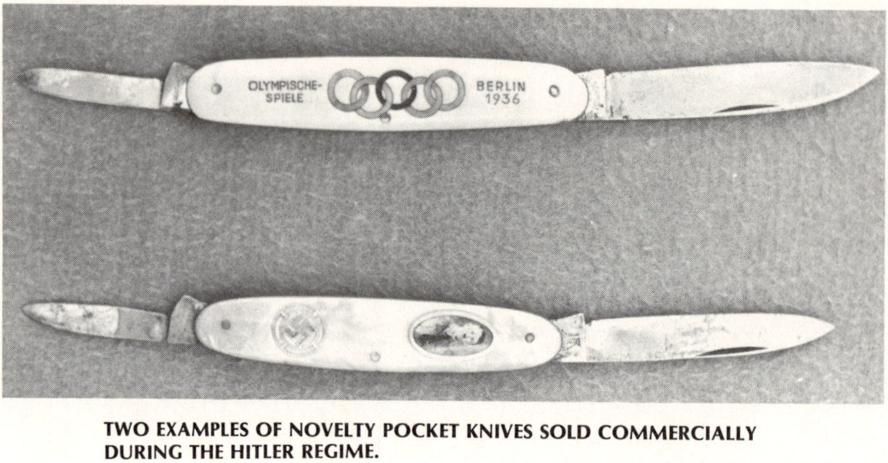

PHOTO: MARC BROWN/HICKOX COLLECTION

TWO EXAMPLES OF NOVELTY POCKET KNIVES SOLD COMMERCIALLY DURING THE HITLER REGIME.

FIVE EXAMPLES OF BLADEWARE PRODUCED BY THE SOLINGEN BLADE FIRMS TO MEET THE DEMAND OF THE COMMERCIAL MARKET. THESE TYPES OF KNIVES WERE USED SIMPLY AS POCKET PIECES OR FOR USE AS A HUNTING KNIFE. MAKERS FROM RIGHT TO LEFT: SMF SOLINGEN, ROBERT KLAAS, NO MAKER, UNKNOWN MAKER, NO MAKER.

ROBERT BOLTON

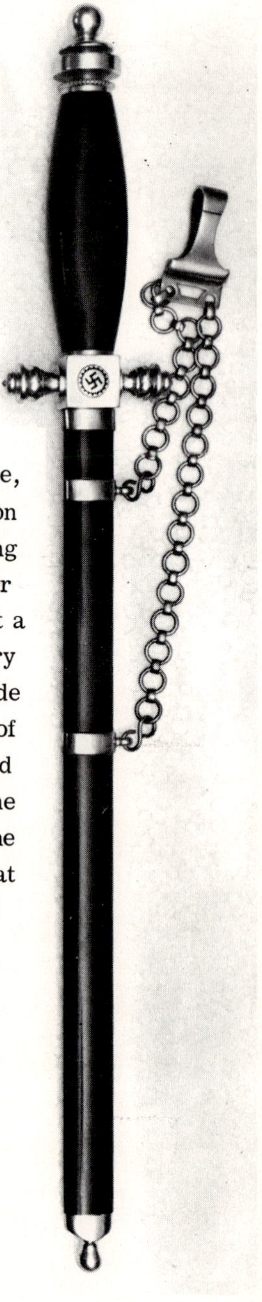

This very interesting sidearm is not a dagger per se, but takes on every aspect of one. Its outer configuration would give the viewer the impression that he was looking at a unique dagger representative of the German Labor Front. However, the blade is not a blade at all, but a steel blade sharpener identical to those used in most every kitchen to sharpen the butcher knife. This piece was made by the firm of F. Dick, not a Solingen-based producer of blades. All metal fittings are gold colored. The grip and the dark portion of the scabbard are a purple color. The insigne of the German Labor Front (DAF) is inlaid on the obverse of the crossguard, while the insigne of the meat cutters craft guild is inlaid on the reverse.

Bayonets

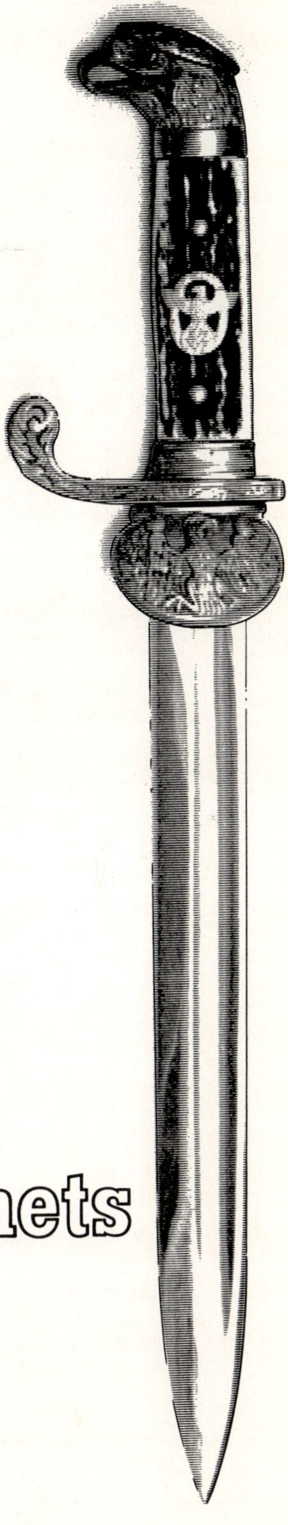

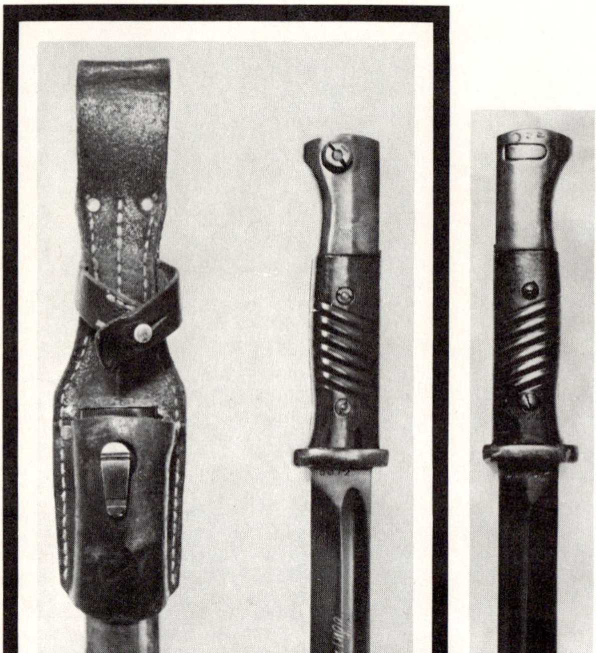

PHOTO: ANDY WALKER

STANDARD MAUSER SERVICE BAYONET BEARING THE SS MOTTO, "MEINE EHRE HEISST TREUE", AND, SS ADOLF HITLER SS", ON THE OBVERSE OF THE BLADE. THE REVERSE BEARS THE OWNER'S NAME. NOT DISCERNIBLE ON THE POMMEL AND RICASSO IS THE MARK OF THE SS AND RZM IN VERY SMALL CIRCLES.

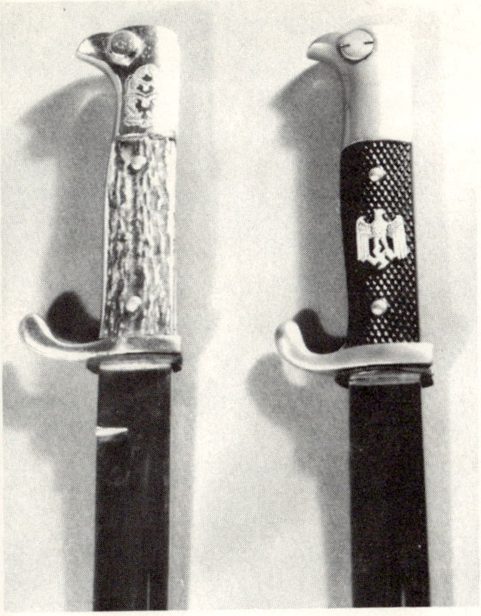

ROBERT BOLTON

TWO PATTERN DRESS SIDEARMS. THE ONE TO THE LEFT HAS A SHORT BLADE WITH STAG GRIPS, AND THE ENGRAVED LUFTWAFFE INSIGNIA ON THE OBVERSE OF THE POMMEL. MANUFACTURED BY EICKHORN. THE SIDEARM TO THE RIGHT IS A BASIC PATTERN ALSO PRODUCED BY THE EICKHORN FIRM, BUT HAVING THE WEHRMACHT EAGLE ON THE OBVERSE GRIP PLATE.

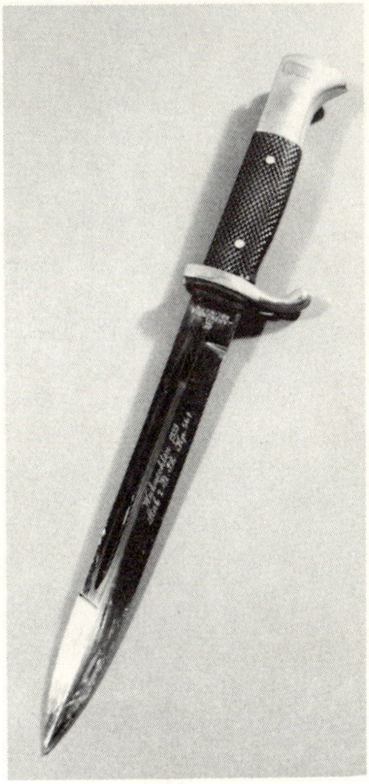

ROBERT BOLTON

NCO DRESS SIDEARM PRODUCED BY J. A. HENCKELS, BEARING THE BLADE DEDICATION, WEIHNACHTEN 1939" (CHRISTMAS 1939), "STAB 2 PI. PK. KP. 561".

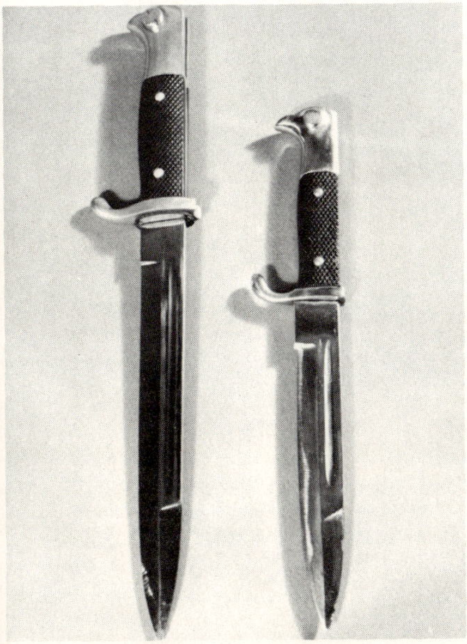

ROBERT BOLTON

THE DRESS SIDEARM AT THE RIGHT IS A FULL-SCALE MINIATURE PRODUCED BY THE PUMA FIRM . . . COMPLETE WITH MOVEABLE LUG FASTENER. THE SIDEARM AT THE LEFT IS FOR COMPARISON PURPOSES ONLY.

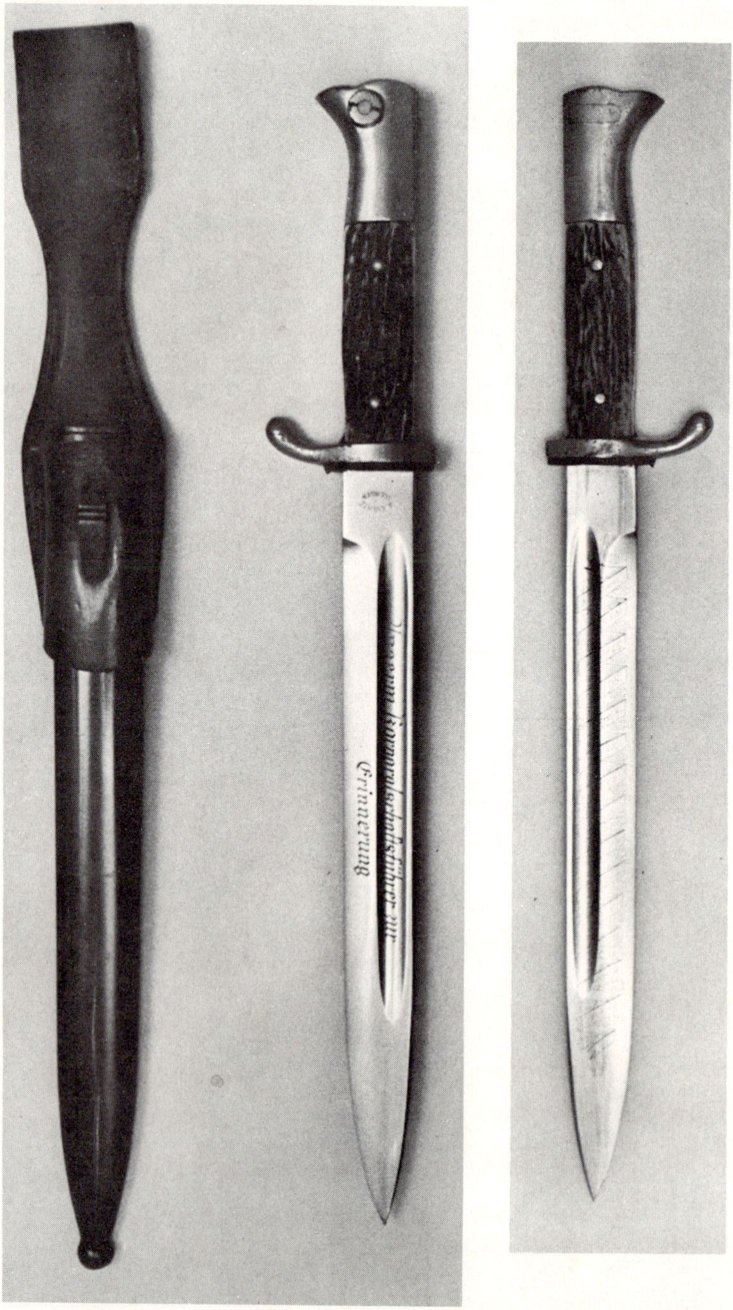

BOB SETTLE

PERSONALIZED DRESS SIDEARM PRESENTED TO THE UNIT SQUAD LEADER, PROBABLY BY MEMBERS OF HIS SQUAD. THE OBVERSE BEARS THE DEDICATION, WHILE THE REVERSE BEARS THE NAMES OF EACH OF THE MEMBERS. THE SIDEARM IS A BASIC PATTERN WITH THE DEDICATION BEING SPECIALLY ORDERED AT TIME OF PURCHASE. THIS WAS A COMMON PRACTICE AND THE DEDICATIONS CAN BE ENDLESS.

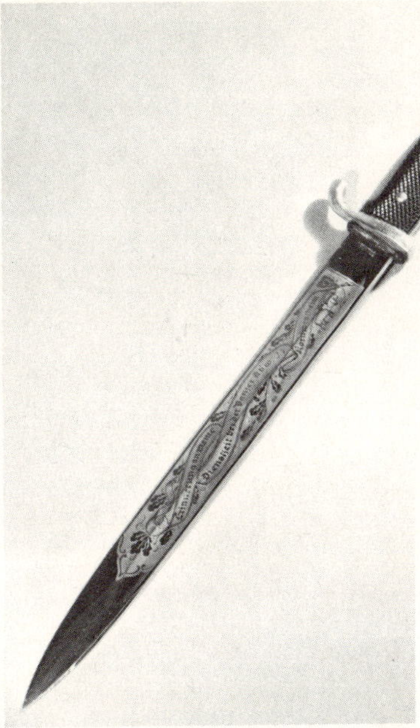

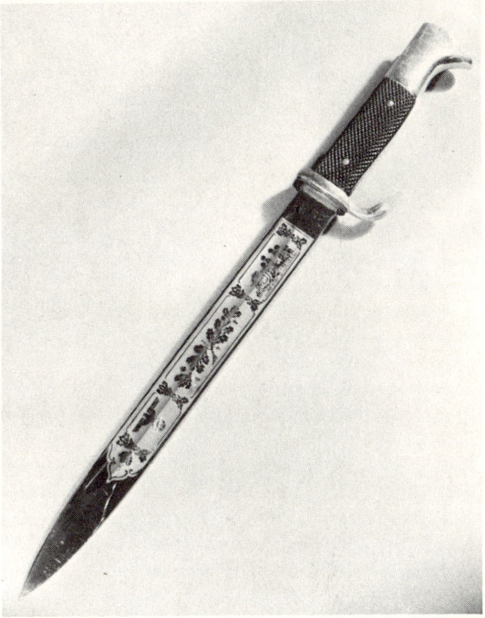

LONG BLADED, BASIC PATTERN DRESS SIDEARM WITH DOUBLE ETCHED PATTERN. LEFT IS THE OBVERSE PATTERN BEARING THE DEDICATION, "ERINNERUNG AN MEINE DIENSTZEIT BEIDER PANZER ABW. 25, KORNWESTHEIM" (IN REMEMBERANCE OF MY SERVICE TIME WITH THE PANZER BATTALION 25, KORNWESTHEIM (LOCATION)). THE RIGHT PHOTO SHOWS THE OBVERSE OF THE BLADE.

MEMBERS OF THE FLA-ERS-BTL (MOT.) 66 SWEAR THE OATH OF ALLEGIANCE AT THEIR HOME STATION AT WACKERNHEIM/ MAINZ. THEY WEAR THE STANDARD PATTERN MAUSER SERVICE BAYONET WITH THE PARADE UNIFORM.

(ABOVE) UNTEROFFIZIER WALTER MARSCHAUS AT THE RIGHT AND HIS FRIEND, BOTH MEMBERS OF THE 10. PANZER-DIVISION, PANZERREGIMENT 7, WEAR THE DRESS SIDEARM WITH PORTEPEE WITH THE WALKING-OUT DRESS UNIFORM.

BOB SETTLE

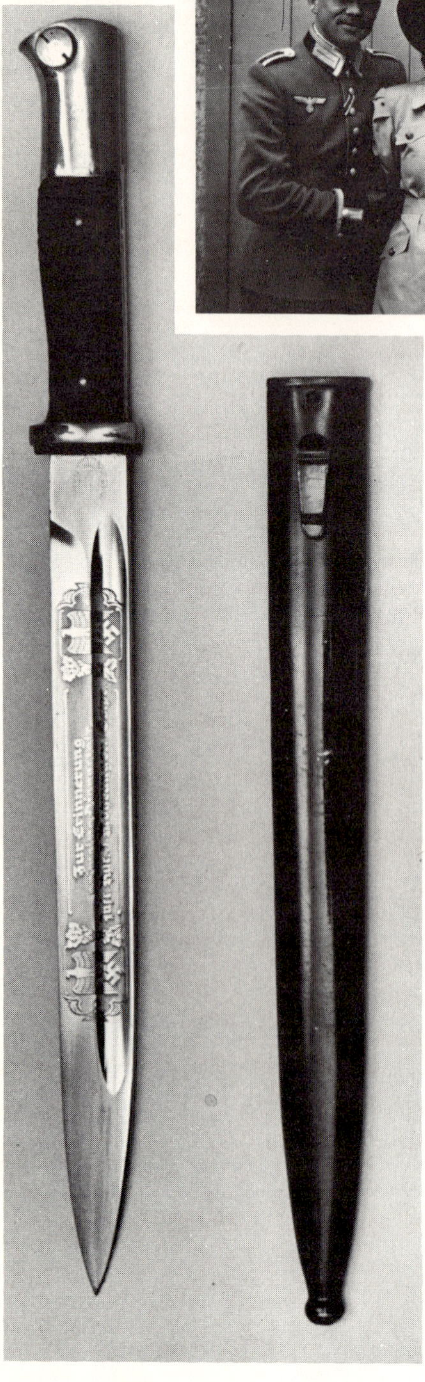

DRESS SIDEARM ATTRIBUTED TO SOME MANUFACTURING FIRMS AS BELONGING SOLELY TO THE PIONEER (ENGINEER) UNITS. HOWEVER, THE ETCHED DEDICATION WOULD INDICATE THAT THE MEMBER IS AN INFANTRYMAN RATHER THAN AN ENGINEER. PRODUCED BY F. W. HOLLER.

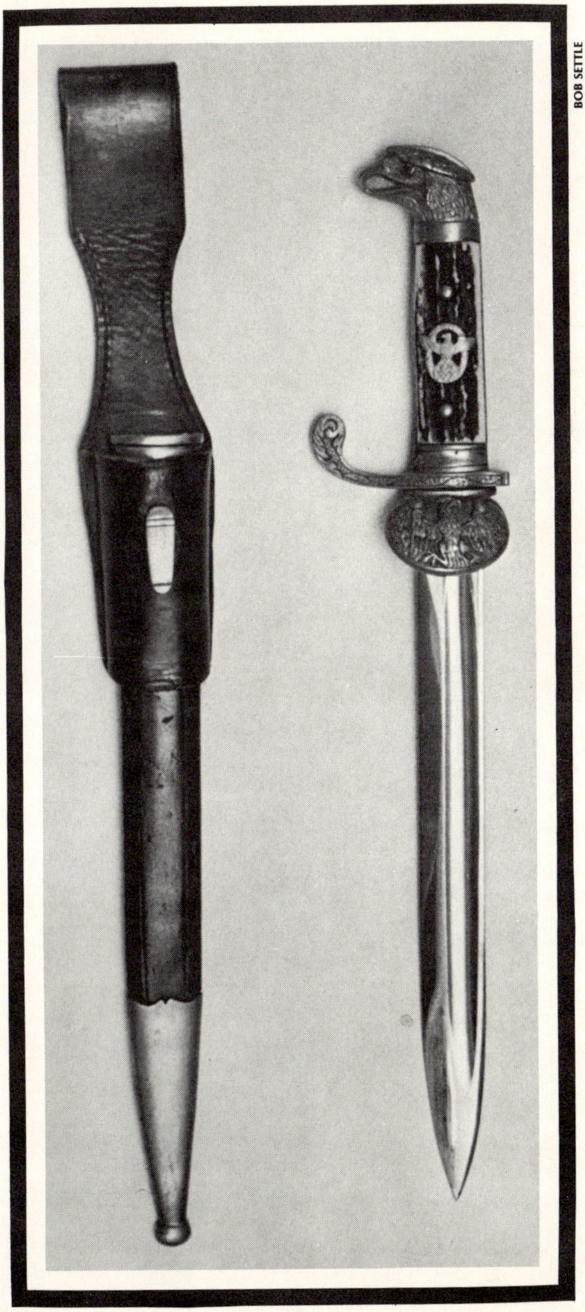

BOB SETTLE

WEIMAR PATTERN DRESS SIDEARM FOR POLICE OFFICIALS WHICH HAS UNDERGONE THE TRANSITION TO THE NAZI REGIME WITH THE ADDITION OF THE POLICE NATIONAL EMBLEM.

Illustrated is a drastically modified Model 98/05 bayonet. The blade has been shortened approximately 14 inches, narrowed by 6mm and nickel-plated. The grip underwent the greatest change with bone grips replacing the wood side-plates, and the Police national emblem added to the front center. Hilt mounts are ornately designed with an engraved pattern and nickel-plated. This modification was worn as late as 1941, and in particular by some members of the SS-Police security units.[1]

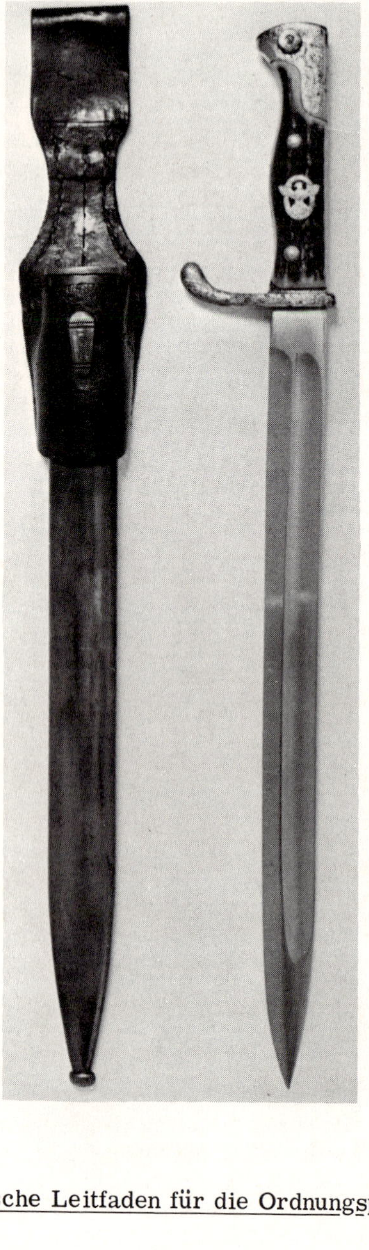
PHOTO: ANDY WALKER

---

[1] Waffentechnische Leitfaden für die Ordnungspolizei, Karl Fischer, 1941.

This Feldjäger dress sidearm shows the first pattern grip-plate insigne without the swastika. This was later changed with slight modification to the design of the eagle to incorporate the swastika. Service patterns of the sidearm are larger in size, with longer blades. The Feldjägerkorps was a short-lived, political police-style organization which was only in existence from 1933, to its incorporation into the Police in 1935. It is interesting to note that, when presenting arms during

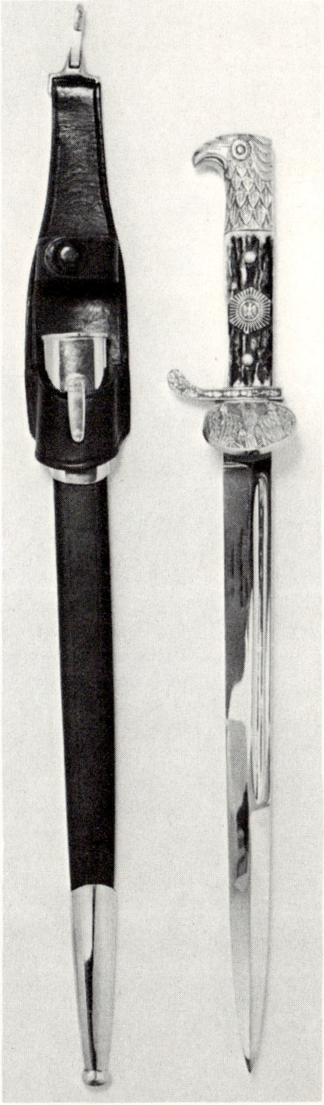

PHOTO: ANDY WALKER

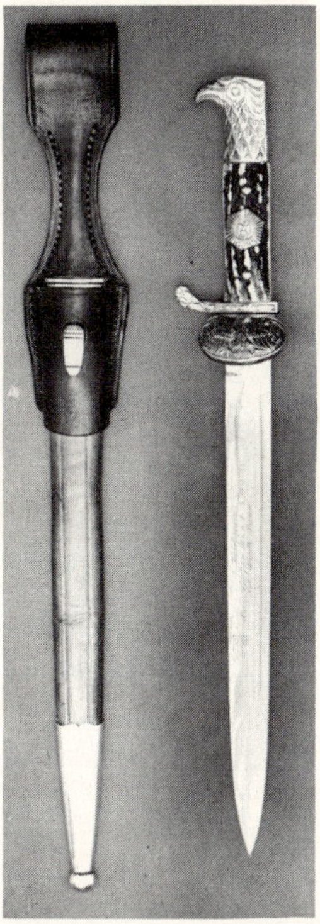

GARY WALKER MUSEUM COLLECTION

FELDJÄGER DRESS SIDEARM PRODUCED BY THE EICKHORN FIRM, AND ENGRAVED FOR PRESENTATION. THE DEDICATION INDICATES THAT THE PIECE WAS A 1st PLACE SHOOTING AWARD TROPHY, PROBABLY AT THE POLICE TRAINING SCHOOL AT BONN.

a formal formation, the bayonet was drawn and held diagonally across the chest... a most unusual practice, but rather significant considering the emphasis placed on the sidearm.

"KÖLNISCHE ILLUSTRIERTE ZEITUNG"

(ABOVE) POLIZEIWACHTMEISTER OF THE KÖNIGSBERGER SCHLOSS (BERLIN) SCHUTZPOLIZEI ON TRAFFIC DUTY. AS LATE AS 1936, ALL MEMBERS OF THE POLICE HAD NOT MADE THE TRANSITION TO THE NEW NAZI UNIFORM FROM THE OLD PATTERN WEIMAR.

AS LATE AS 1938, THIS SA STANDARD BEARER CONTINUED TO WEAR A BAYONET SIDEARM. THE PIECE IS BELIEVED TO BE OF AUSTRIAN MANUFACTURE AND DESIGN.

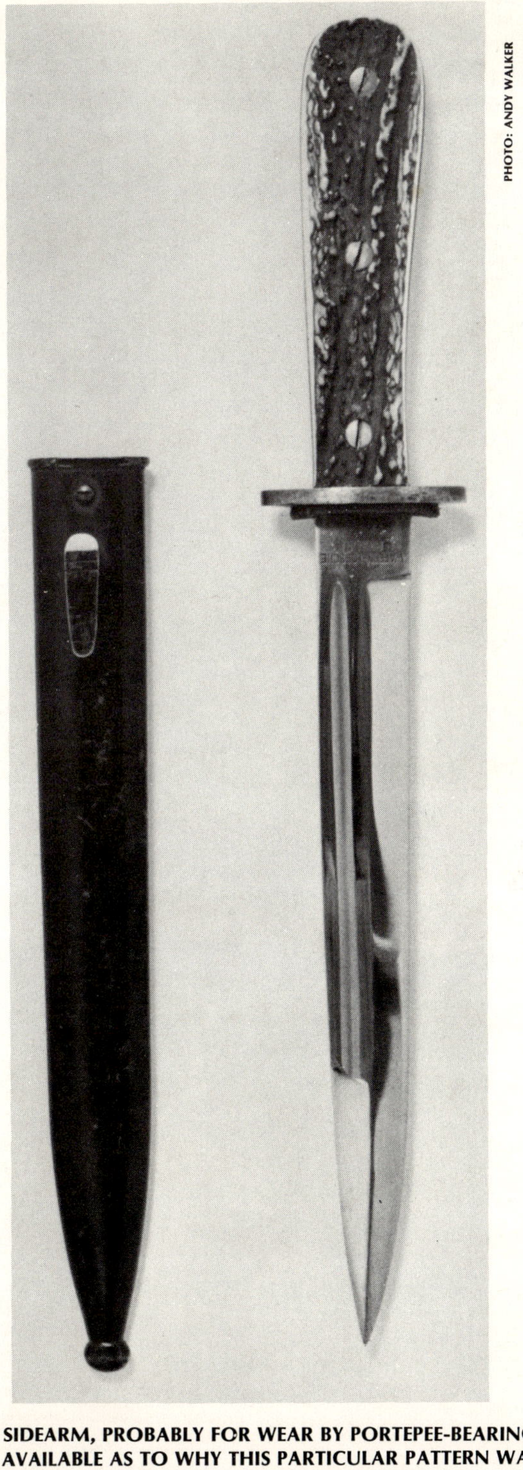

LUFTWAFFE DRESS SIDEARM, PROBABLY FOR WEAR BY PORTEPEE-BEARING NCO's. LITTLE INFORMATION IS AVAILABLE AS TO WHY THIS PARTICULAR PATTERN WAS WORN BY THE LUFTWAFFE OR WHY THE CHANGE IN PATTERN FROM THE REGULATION DRESS SIDEARM. "WAFFEN-LOESCHE BERLIN" IS STAMPED ON THE OBVERSE WHILE "CH A. W." IS STAMPED ON THE REVERSE OF THE RICASSO.

MAJ. TOM JOHNSON

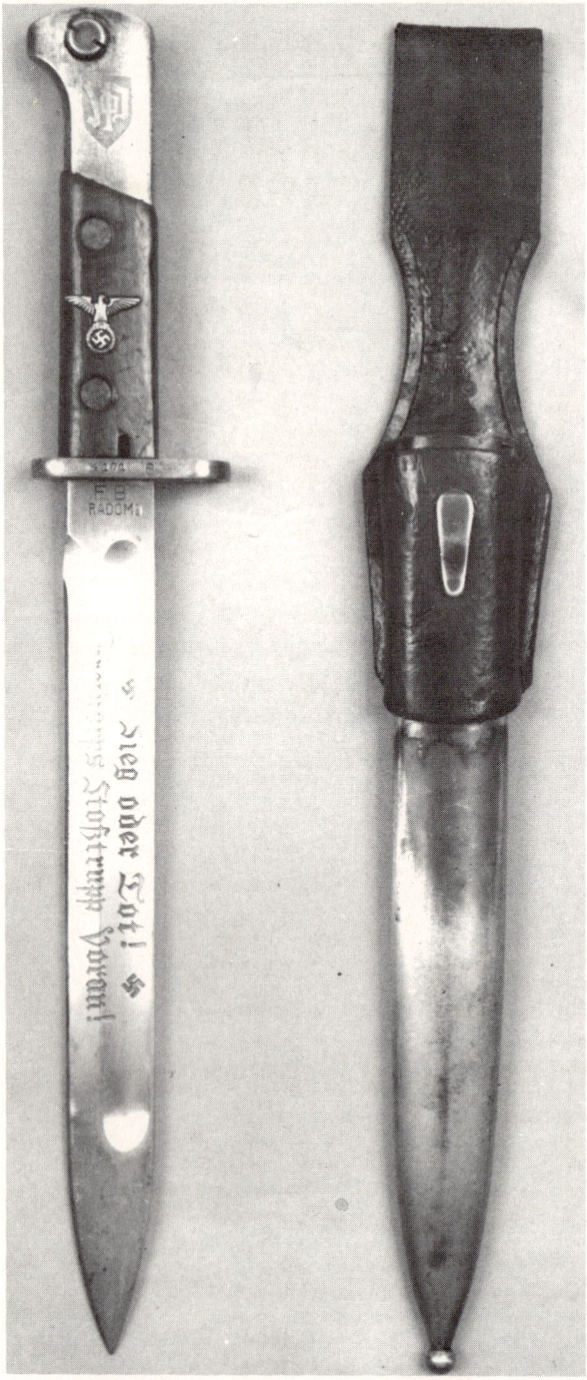

PERSONALIZED AUSTRIAN BAYONET MODEL 1888 BEARING THE MONOGRAM "SdP" (SUDETENDEUTSCHE PARTEI) ON THE POMMEL, AND ETCHED BLADE INDICATING THE BEARER WAS A MEMBER OF THE FREIKORPS STROSSTRUPP "VORAN". THE SILVER NATIONAL EMBLEM INLAID IN THE WOOD GRIP-PLATE IS THE EARLY PATTERN NSDAP EAGLE. THE "RADOM" MARKING ON THE LANGET IS THE MANUFACTURER'S DESIGNATION.

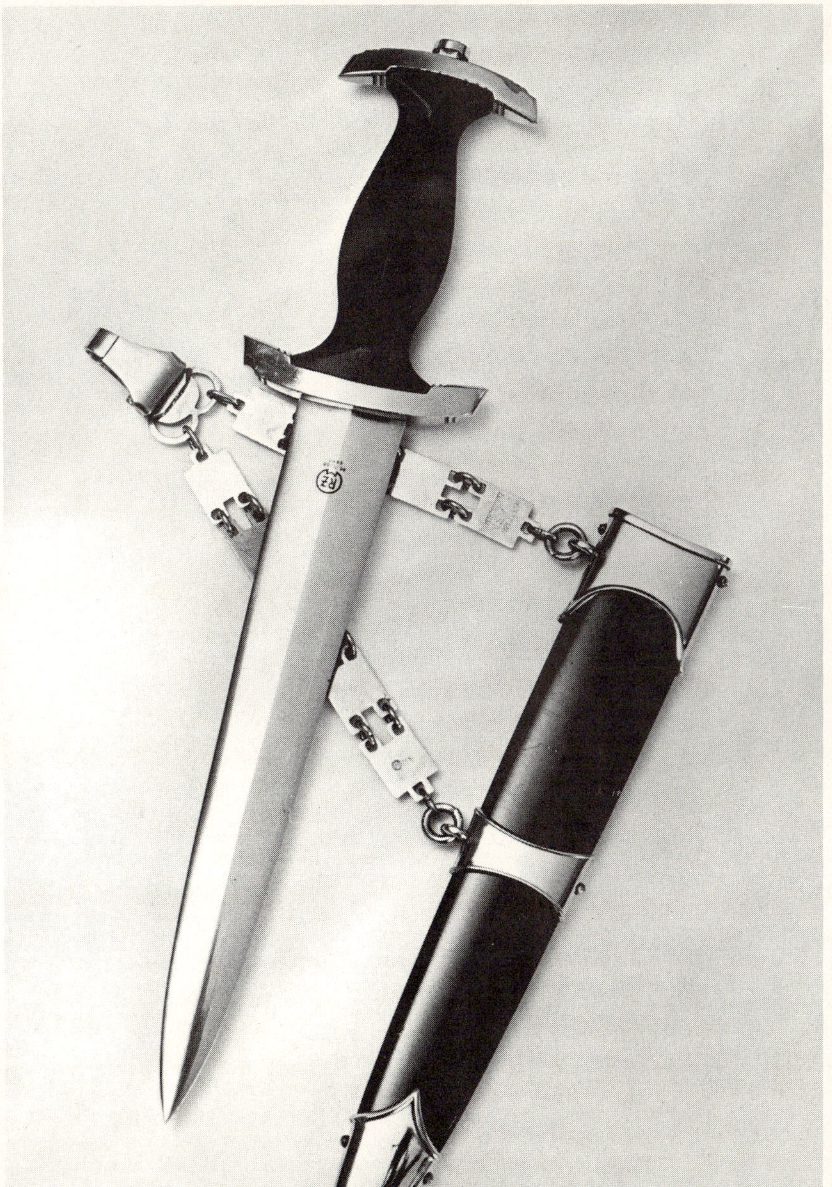

RZM Markings, etc.

DAGGER MARKINGS

At the time "Daggers, Bayonets and Fighting Knives of Hitler's Germany" was published, it was recognized that the chapter dealing with the manufacturer's RZM control codes was far from complete. A request for unlisted codes which could be positively identified with the manufacturer was made, and the response was great. It is as a result of the reader's interest that these additional codes follow. Not wishing to duplicate that which has already been written, use as a base reference the codes shown on pages 314-316 of the above reference.

RZM M7/16	Justus Brenger & Co., Solingen-Wald
RZM M7/20	Ernst Mandewirth
RZM M7/28	G. Felix, Gloriawerk, Solingen
RZM M7/32	Robt. Müller, Solingen-Merscheid
RZM M7/44	F.W. Backhous
RZM M7/47	Paul Ebel, Solingen
RZM M7/63	Herder & Engels, Solingen-Ohligs
RZM M7/65	Carl Heidleberg, Solingen
RZM M7/77	Gustav Schmeider
RZM M7/87	A.M. Bronn
RZM M7/90	Karl Teigel
RZM M7/91	Carl Spitzer Malsch
RZM M7/93	Ewald Cleff
RZM M7/96	Drees & Sohn
RZM M7/105	Rudolf Wurzer
RZM M7/106	Georg Kerschbaumer
RZM M7/163	E & A Helbig, Steinbach
RZM M7/183	Ernst E. Witte (Kroneck), Solingen
RZM M7/1053	P.D. Lüneschlos, Solingen (see M7/14)
RZM M7/1164	David Malsch, Steinbach Krm.
RZM M7/1211	E. Pack & Söhne, Solingen

Obviously, there are yet more to be identified. Many code designations were encountered, but no positive fix could be established with the manufacturer.

Some interesting observations have been derived as a result of studying the RZM code system. First, on rare occasions, duplicate numbers were assigned to different manufacturers. An example of this is RZM M7/29 being assigned to both Klittermann & Moog, and also Jacobs & Co. Secondly, in the case of four digit numbers, it would apprear that these are additional codes given to firms with a previously assigned RZM code designation. An example of this is as shown above with the firm of P.D. Lüneschlos, assigned RZM M7/14 and M7/1053.

As firms were assigned code designations, the details pertaining to the firm along with the code assigned, were published in the "RZM Mitteilungsblatt". This was followed with a periodic up-dating of firms authorized to produce the Hitler Youth or political form daggers, and published in the trade journal "Die Klinge".

ORDNANCE MARKS

The ordnance inspection mark or "Waffen-Amt" was the inspectors' control mark indicating that the piece had met with quality control requirments. While inspection marks were placed on many military or military related blades, only the "Waffen-Amt" is found on military combat sidearms. The "Waffen-Amt" underwent two changes in 1937, going from the old pattern to one simulating the Nazi-style national emblem:

TO

TO

It was common practice to recondition and issue captured stocks to combat troops. In such a case, a "Waffen-Amt" could be found on a foreign manufactured bayonet. In addition, some firms in occupied countries were required to continue production for the German fighting man. In this case as well, the "Waffen-Amt" was applied to the pommel, scabbard, and sometimes blade.

Inspection control marks of the Kriegsmarine (KM) are as follows:

(WEIMAR)

(PRE-1938)

(POST-1938)

RUNIC SYMBOLS

During the Hitler era, extensive study was given to the runic symbols... largely because the Nazis took great pains to trace their beginnings to the Nordic tribes. Specific letters of the runic alphabet were incorporated into the design of some of the daggers, or complete dedications from the Futhark or runic language were inscribed on the blades, using solely the runic alphabet. The runic alphabet consisted of between 16 to 27 letters. Some professors working with the subject have even gone as far as attributing the Roman alphabet to the earlier established runic alphabet. Using a backward progression from the early forms of the runic alphabet, Professor Hermann Wirths, in particular, was able to establish the origins of the Roman alphabet as follows:

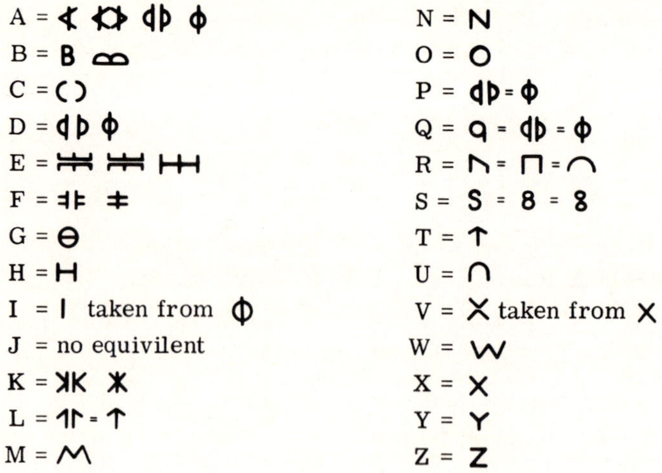

The logic used by Professor Wirths to arrive at his conclusion stemmed from a progression from the "sun sign", which in turn, was derived from the basic circle (representing the sun) and the X (representing the points of the compass). The sun sign was broken into eight basic groupings, each with a direct relation to nature or life. From this, a rather complicated transformation was established, and Professor Wirths and his colleagues were able to "prove" the early "origins" of the Roman alphabet.

The runic alphabet to be used by the engravers and blade producers of the Nazi period, however, was not that shown above, but was a less complicated form as shown here:[1]

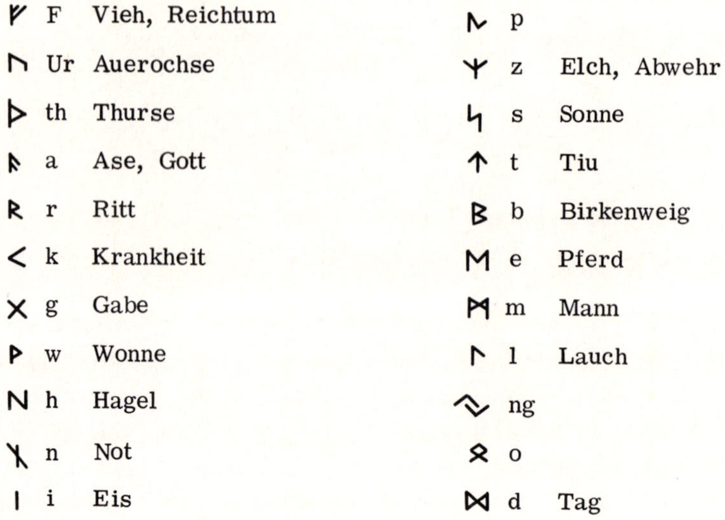

---

[1] SS-Leitheft, Heft 4, 1942.

The preceding runic alphabet is not intended to be in alphabetical order, but in the order established and accepted by scholars during the Nazi period.

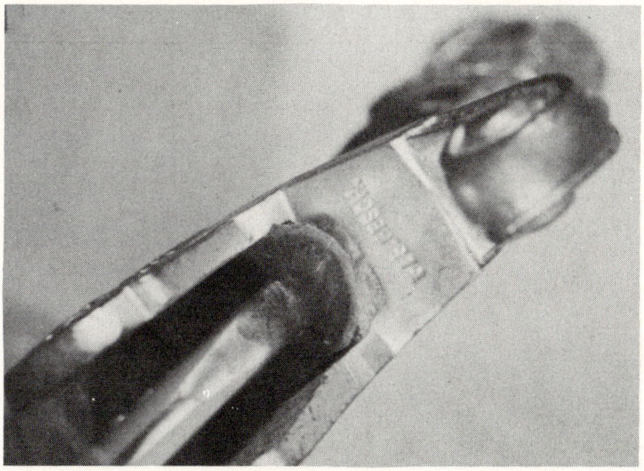

Illustrated is the mark, "Ges. Gesch." (gesetzlich geschützt) which in its literal sense means patent pending. This mark, shown on the underside of a sword hilt, indicates that the manufacturing firm has a design or some feature of the manufacturing process pending patent approval.

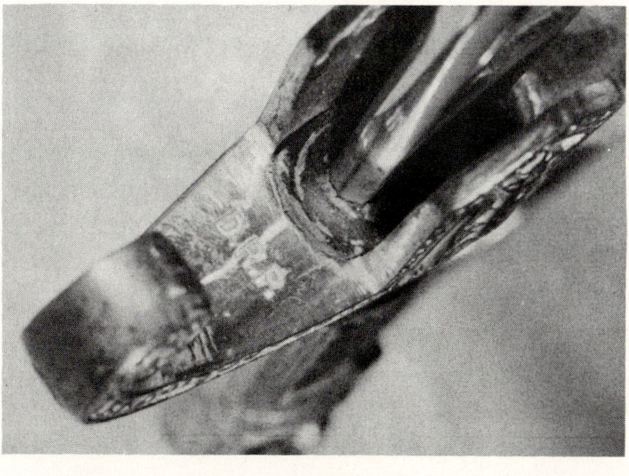

This D. R. P. designation (Deutsches Reichs Patent) on the underside of a sword hilt indicates that the patent has been approved, and the design or manufacturing technique cannot be copied by another firm without prior approval. This mark is not to be confused with the DRP or "Deutsches Reichs Post" abbreviation which is normally found on the Postal daggers and bayonets.

A periodic listing of firms authorized by the RZM to produce the Hitler Youth Knife and SA daggers was published in the monthly trade journal, "Die Klinge". A study of successive listings shows that such authorization was sometimes withdrawn. The trademarks shown are representative of the firms, but were subject to change through the years. This particular listing was taken from the April 1940 edition of "Die Klinge". This and other data extracted from the journals were made available through the courtesy of the Deutsches Klingenmuseum, Solingen, and Dr. Hanns-Ulrich Haedeke, Director of the Museum.

Hersteller von SA-Dolchen und HJ-Fahrtenmessern

Zugelassen von der Reichszeugmeisterei München

	Gebr. Bell Solingen-Gräfrath M 7/94	Karl Böcker Solingen, Schützenstr. 24 M 7/45
Gebr. Born Solingen M 7/82	GEBR. CHRISTIANS CHRISTIANSWERK Solingen M 7/1	Carl Eickhorn Solingen M 7/66
Arthur Evertz Solingen M 7/85	Gebr. Gräfrath Solingen-Widdert M 7/30	Josef Hack Ges. m. b. H. Steyr (Ober-Donau) M 7/103
Herm. Hahn Solingen-Wald M 7/71	Carl Halbach Solingen M 7/23	Gottlieb Hammesfahr Solingen-Foche M 7/67
J. A. HENCKELS • ZWILLINGSWERK • SOLINGEN M 7/10		
Hartkopf & Co. Solingen M 7/40	Herbertz & Meurer Solingen-Gräfrath M 7/52	Friedr. Herder A. S. Constantwerk Solingen M 7/49
H. HERDER Solingen M 7/111	Richard Abr. Herder Solingen M 7/18	Robert Herder Solingen-Ohligs M 7/55
F. W. Höller Solingen M 7/33	E. & F. Hörster Solingen M 7/36	Juliuswerk J. Schmidt & Söhne Riemberg i. Schlesien M 7/88
Karl Rob. Kaldenbach Solingen-Gräfrath M 7/72	Robert Klaas Solingen-Ohligs M 7/37	Klittermann & Moog G. m. b. H. Haan bei Solingen M 7/29
E. Knecht & Co. Solingen M 7/11	F. Koeller & Co. Solingen-Ohligs M 7/97	Gustav L. Köller Nachf. Solingen-Wald M 7/60
Herm. Konejung A.-G. Solingen M 7/7	Carl Julius Krebs Solingen M 7/5	Pet. Dan. Krebs Solingen M 7/92
Lauterjung & Co. Tiger-Stahlwaren u. Waffenfabrik Solingen M 7/68	Carl & Robert Linder Solingen-Weyer M 7/15	Herm. Linder Söhne Solingen M 7/78
Hugo Linder C. W. Sohn Solingen-Weyer M 7/114	P. D. Lüneschloß Solingen M 7/14	August Merten Ww. Solingen-Gräfrath M 7/31
Friedr. Plücker jr. Solingen-Gräfrath M 7/62	Puma-Werk Lauterjung & Sohn Solingen M 7/27	Cuno Remscheid & Co. Solingen-Aufderhöhe M 7/86
Kuno Ritter Solingen-Gräfrath M 7/3	Solinger Metallwaren- Fabrik Stöcker & Co. Solingen M 7/9	C. Gustav Spitzer K.-G. Solingen M 7/80
C. D. Schaaff Solingen M 7/56	Carl Schmidt Sohn K.-G. Solingen M 7/84	J. A. Schmidt & Söhne Solingen M 7/95
Rudolf Schmidt Solingen M 7/41	Artur Schüttelhöfer & Co. Solingen-Wald M 7/13	Friedrich Aug. Schmitz Solingen M 7/74
Karl Tiegel Riemberg Bez. Breslau M 7/81	Emil Voos Solingen M 7/2	WAFFENFABRIK MAX WEYERSBERG Solingen M 7/12
Wilh. Wagner Solingen-Merscheid M 7/25	Paul Weyersberg & Co. Solingen M 7/43	Ernst Erich Witte Solingen M 7/98
WKC-Waffenfabrik Gesellschaft m. beschränkt. Haftung Solingen-Wald M 7/42	Carl Wüsthof Gladiatorwerk Solingen M 7/112	Ed. Wüsthof Dreizackwerk Solingen M 7/19

BIBLIOGRAPHY

Angolia, John R.: Daggers, Bayonets & Fighting Knives of Hitler's Third Reich, Calif., 1971.

Angolia, John R.: Swords of Hitler's Third Reich, England, 1969.

Atwood, James P.: The Daggers and Edged Weapons of Hitler's Germany, Omnium-Druck und Verlag, Germany, 1965.

Bekleidungs- und Anzugsbestimmungen für die Kriegsmarine M. Dv. Nr. 260, Berlin, 1935.

Bender, Roger James,: Luftwaffe, Calif., 1972.

Blanke Waffen, Solingen, 1941.

Denckler, Heinz: Pimpf, dein Dienst im Deutschen Jungvolk in der Hitler-Jugend, Berlin.

Die Seitenwaffen, Manual for Ordnance and Supply Personnel, as of April 12, 1940, Berlin, 1941.

Eickhorn Kundendienst, Sales catalogue for the Eickhorn firm, 1939, 1941.

German Army Combat and Construction Engineer Manual, Berlin, 1937.

German Military Uniforms and Insignia 1933-1945, We Inc., Ct., 1967.

Göring, Emmy: An der Seite meines Mannes, Göttingen, 1967.

Gritzbach, Erich: Hermann Göring, Werk und Mensch, Zentralverlag der NSDAP, 1943.

Handbook on German Military Forces - TM-E 30-451, US Army, 1945.

Haid, Georg: Ein Buch vom neuen Heer, Stuttgart, 1935.

Die Klinge, Monthly Trade Magazine, Solingen, 1933-1942.

Klietmann, Dr. Kurt-Gerhard: German Daggers and Dress Sidearms of World War II, Falls Church, Va., 1967.

Manvell, Roger: Göring, Ballantine Books, N.Y., 1972.

Mollo, Andrew: Daggers of the Third German Reich - 1933-1945, London, 1967.

Mollo, Andrew: Uniforms of the SS, Vols. 1-6, London.

Organisationsbuch der NSDAP, 1937-1943 editions.

Sales Catalogues of:
- ALCOSO, 1941
- Herder, 1938
- Höller, 1938, 1941
- Grafarth, 1938
- Klaas,
- Krebs, 1938
- Puma, 1940
- Seilheimer, 1938
- WKC, 1939

Shirer, William L.: The Rise and Fall of the Third Reich....A History of Nazi Germany, Simon and Schuster, N.Y. 1960.

SS-Dienstaltersliste der Schutzstaffel der NSDAP, Organization List of SS Officer Personnel, Stand vom 1 Dezember 1937.

Stephens, Frederick J.: Edged Weapons of the Third Reich, 1933-1945, England, 1972.

Uniformen-Markt, Trade Paper of the Manufacturers, selected editions.

Vagts, Alfred: Hitler's Second Army, Infantry Journal, Washington, D.C., 1943.

Wilkinson, Frederick: Swords & Daggers, England, 1967.